T0173128

International Mathematics

Coursebook 2

Andrew Sherratt

HODDER
EDUCATION
PART OF HACHETTE LIVRE UK

Hachette UK's policy is to use papers that are natural, renewable and recyclable products and made from wood grown in well-managed forests and other controlled sources. The logging and manufacturing processes are expected to conform to the environmental regulations of the country of origin.

Orders: please contact Hachette UK Distribution, Hely Hutchinson Centre, Milton Road, Didcot, Oxfordshire, OX11 7HH. Telephone: +44 (0)1235 827827. Email education@hachette.co.uk Lines are open from 9 a.m. to 5 p.m., Monday to Friday. You can also order through our website: www.hoddereducation.co.uk

© Andrew Sherratt 2009
First published in 2009 by
Hodder Education, an Hachette UK Company.
338 Euston Road
London NW1 3BH

Impression number   10
Year   2022

All rights reserved. Apart from any use permitted under UK copyright law, no part of this publication may be reproduced or transmitted in any form or by any means, electronic or mechanical, including photocopying and recording, or held within any information storage and retrieval system, without permission in writing from the publisher or under licence from the Copyright Licensing Agency Limited. Further details of such licences (for reprographic reproduction) may be obtained from the Copyright Licensing Agency Limited, www.cla.co.uk

Cover photo © Michael Betts/Photographer's Choice/Getty Images
Illustrations by Macmillan Publishing Solutions and Robert Hichens Designs
Typeset in 12.5pt/15.5pt Garamond by Macmillan Publishing Solutions
Printed by CPI Group (UK) Ltd, Croydon CR0 4YY

A catalogue record for this title is available from the British Library

ISBN 978 0 340 96743 0

Contents

Unit 1

Measurement of plane figures — **1**
A Systems of measurement — 1
B What is a plane figure? — 4
C Some special plane figures — 5
D Perimeter — 7
E Perimeter of a rectangle, square parallelogram and rhombus — 8
F Area — 10
G Area of rectangles and squares — 10
H Area of a parallelogram — 16
I Area of a triangle — 18
J Area of a trapezium — 22
K Working out the area of compound figures — 24
L Circles — 27

Unit 2

Algebraic equations in one variable — **33**
A What is an equation? — 33
B Solving equations by inspection — 33
C Solving equations by working backwards — 35
D Solving equations using the 'balance method' — 37
E Writing equations to solve word problems — 39

Unit 3

Ratio, proportion and percentage — **48**
A Ratio — 48
B Direct proportion — 58
C Inverse proportion — 62
D Percentage — 64

Unit 4

Applications of ratio, proportion and percentage — **77**
A Ratio — 77
B Proportion — 78
C Percentage — 79
D Simple financial calculations — 82

Unit 5

Square roots and cube roots — **103**
A Square roots — 103
B Cube roots — 112
C Some properties of square roots — 115

Unit 6

Real numbers — **118**
A Real numbers — 118
B Rational numbers — 119
C Irrational numbers — 124

Unit 7

Indices — **128**
A Indices — 128
B The third law of indices — 131
C The fourth law of indices — 133
D The fifth law of indices — 134
E Negative indices — 136
F Fractional indices — 141
G Using the laws of indices to solve equations — 146

Unit 8 | **Algebraic multiplication and division** | **149**
A Expansion using the distributive law | 149
B Expansion of two expressions in the form $(a + b)(c + d)$ | 151
C Expansion of two expressions in the form $(a + b)(c + d + e)$ | 153
D Some special products of expansion | 154
E Using the rules of algebra to make calculations easier | 157
F Division of algebraic expressions | 158

Unit 9 | **Unit 9 Factorising quadratic expressions** | **163**
A Expansion of algebraic expressions | 163
B Factorisation of quadratic expressions | 164

Unit 10 | **Unit 10 Pythagoras' theorem** | **181**
A Who was Pythagoras and what is a theorem? | 181
B Pythagoras' theorem | 181
C Finding the hypotenuse | 182
D Finding one of the shorter sides | 183
E Using Pythagoras' theorem to solve word problems | 186
F Solving problems in three dimensions | 190
G Using the converse of Pythagoras' theorem | 194

Unit 11 | **Parallel lines, triangles and polygons** | **198**
A Some angle facts | 198
B Parallel lines | 198
C Triangles | 208
D Polygons | 223

Unit 12 | **Patterns and sequences** | **227**
A What is a sequence? | 227
B Using a rule to write the next term in a sequence | 228
C Using a rule to write any term in a linear sequence | 232
D Finding a rule to write any term in a linear sequence | 233
E Number sequences from shape patterns | 239
F Some special number sequences | 240

Unit 13 | **Presenting statistical data** | **245**
A Data and statistics | 245
B Line graphs | 246
C Pictograms | 246
D Bar charts | 247
E Frequency tables | 248
F Histograms | 249
G Stem-and-leaf diagrams | 249
H Pie charts | 251
I Collecting data through interviews | 259

Unit 1 Measurement of plane figures

Key vocabulary

arc	inch	quadrilateral
area	isosceles triangle	radius
circle	litre	rectangle
circumference	metre	rhombus
compound figure	metric system	right-angled triangle
diameter	ounce	scalene triangle
equilateral triangle	parallel	square
foot	parallelogram	stone
gallon	perimeter	trapezium
gram	pint	unit
hectare	plane figure	yard
Imperial system	pound	

A Systems of measurement

Revision

We have already learned that we use a **measure** to tell us 'how much' we have of something. In the past, people used many different ways to measure things – for example, the length of a man's thumb, or his foot, or a step he can walk, have all been used to say how long things are.

I could say that a room is 34 of 'my feet' long; or that my desk is 65 of 'my thumbs' wide. This is fine for me, but everybody else doesn't know how big my foot or my thumb is!

To make things easier, everyone has agreed to use the same references or units to measure different things. The units of measurement that make up the metric system are used all over the world. Another set of units, the Imperial system, is also used in England and some other countries.

The metric system

The **metric system** of units is a **decimal** system − everything is based on multiples of 10 and powers of 10.

Length

In the metric system of measures, the basic unit used to measure length is the metre (m).

1 metre	= 10 decimetres (dm)
1 metre	= 100 centimetres (cm)
1 metre	= 1000 millimetres (mm)
1 decimetre	= $\frac{1}{10}$ metre
1 centimetre	= $\frac{1}{100}$ metre
1 millimetre	= $\frac{1}{1000}$ metre
1000 metres	= 1 kilometre (km)
1 metre	= $\frac{1}{1000}$ kilometre

Mass

In the metric system of measures, the basic unit used to measure mass is the gram (g).

1 gram	= 1000 milligrams (mg)
1 milligram	= $\frac{1}{1000}$ gram
1000 grams	= 1 kilogram (kg)
1 gram	= $\frac{1}{1000}$ kg
1 milligram	= $\frac{1}{1000\,000}$ kg
1000 kilogram	= 1 tonne (t)
1 kilogram	= $\frac{1}{1000}$ tonne
1 gram	= $\frac{1}{1000\,000}$ tonne
1 milligram	= $\frac{1}{1000\,000\,000}$ tonne

Volume

In the metric system of measures, the basic unit used to measure volume is the litre (l).

1 litre	= 1000 millilitres (ml)
1 millilitre	= $\frac{1}{1000}$ litre

NOTE: The abbreviation for 'litre' is 'l', but we usually write the word out in full, because 'l' looks very like the number 1, which can be confusing.

It is very important that you are able to convert one metric unit to a different metric unit (for the same kind of measurement). Practise doing this by revising all the exercises you did last year, and then try this one!

Revision Exercise 1

1 Write down the missing numbers.

a) 320 000 ml = _____ litre

b) 0.32 t = _____ kg = _____ g

c) 3200 g = _____ kg = _____ t

d) 320 mm = _____ cm = _____ m

e) 32 000 cm = _____ m = _____ km

f) 3.2 km = _____ m = _____ cm

2 Which of these lengths are equal?

a) 2000 m
b) 20 km
c) 20 000 cm
d) 2 km
e) 0.02 km

3 Which of these masses are equal?

a) 8000 mg
b) 8 kg
c) 8000 g
d) 0.8 kg
e) 80 kg

4 Which of these lengths is the longest?

a) 0.5 km
b) 50 m
c) 5000 mm
d) 500 cm

5 Which of these masses is the heaviest?

a) 0.3 t
b) 3000 g
c) 3 kg
d) 30 kg

6 A can of coke contains 330 ml. How many litres are there in six cans of coke?

7 One lap of a running track is 400 m. How many laps are run in an 8 km race?

8 There are 20 children at a party. 1 kg of sweets is shared equally among them. How many grams of sweets does each child receive?

9 You need 240 g flour to make 12 biscuits. How many biscuits can you make with 1.2 kg flour?

10 Ben has $\frac{1}{4}$ litre of medicine. He takes 10 ml four times a day for five days. How much medicine does he have left?

The Imperial system

The Imperial units of measurement are still used in some countries. However, it is becoming more common now to use the metric system all over the world.

Length

1 foot (1 ft, 1′) = 12 inches (12 in, 12″)
1 yard (1 yd) = 3 feet

Mass

1 pound (1 lb) = 16 ounces (16 oz)
14 pounds = 1 stone (1 st)

Volume

1 gallon (1 gal) = 8 pints (8 pt)

Converting metric and Imperial measurements

It is sometimes necessary to change Imperial units into metric units (or metric units into Imperial units). For example, we might want to compare measurements made using the two systems.

It is not possible to convert metric and Imperial units exactly, so we use close **approximations**.

Length

1 inch is about 2.5 cm.
1 foot is about 30 cm.
5 miles is about 8 km.

Mass

1 kg is about 2.2 pounds.

Volume

1 litre is about 1.75 pints.
1 litre is about 0.2 gallons.

B What is a plane figure?

In Coursebook 1 we learned that a **plane** has both **width and length**, but no **thickness**. It is like a very big flat surface − imagine a very big piece of paper that goes on forever. A plane is a **flat 2-dimensional surface**.

We use the word **figure** to describe any **shape**.

So a plane figure is a flat shape in two dimensions.

Parallel lines are lines that are **always** the same distance apart. They can **never cross** because they never get closer together (or further apart) no matter how long we make them.

C Some special plane figures

Quadrilaterals

A quadrilateral has four sides and four angles (quad = four and lateral = side).

Rectangle

In a rectangle, the **opposite sides** are **equal** in length, and they are **parallel**. **All the angles** in a rectangle are **90°**.

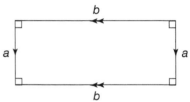

Square

A square is a special kind of rectangle.
All the sides of a square are **equal** in length, and **opposite sides** are **parallel**.
All the angles in a square are **90°**.

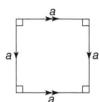

Parallelogram

In a parallelogram, the **opposite sides** are **equal** in length, and they are **parallel**.
The angles are **not** 90°.

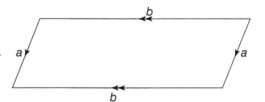

Rhombus

In a rhombus, **all the sides** are **equal** in length, and **opposite sides** are **parallel**.
The angles in a rhombus are **not** 90°.

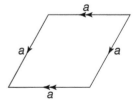

Trapezium

In a trapezium, **one pair of opposite sides** are **parallel**.
All the sides usually have **different lengths**.

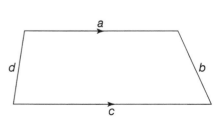

The angles in a trapezium are **not** normally 90° (but some of the angles can be 90° in a special trapezium).

Triangles

All **triangles** have **three sides** and **three angles**. The **sum** of the three angles is always **180°**.

Scalene triangle

In a scalene triangle, the sides all have **different lengths**. The angles are all different.

i) Some scalene triangles have three acute angles.

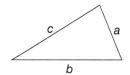

ii) Some scalene triangles have two acute angles and one obtuse angle.

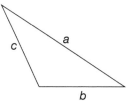

Isosceles triangle

In an isosceles triangle, **two sides** are **equal** in length.
The **two angles** that are **opposite** these equal sides are also **equal**.

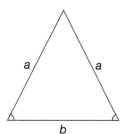

Equilateral triangle

In an equilateral triangle, **all three sides** are **equal** in length.
All **three angles** in an equilateral triangle are **equal** to **60°**.

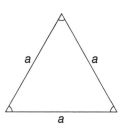

Right-angled triangle

In a right-angled triangle, **one of the angles** is equal to **90°**.
All the sides usually have **different lengths** (but the two sides that make the right angle can sometimes be equal in length).

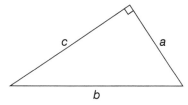

Circles

A circle has just **one side** (or maybe so many that we cannot count them!)
Every point on a circle is the **same distance** from the point called the **centre**.
This distance from the centre to the outside of the circle is called the radius (*r*).
A line segment joining one point on a circle to another point on the circle and **passing through the centre** is called the diameter (*d*).
The length of the diameter is twice the radius ($d = 2 \times r$).
The length around the outside (perimeter) of the circle is called the circumference.
One **part** of the outside of the circle is called an arc.
There are **360°** in a circle.

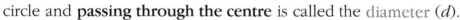

D Perimeter

The **length** (or distance) around the whole of the outside edge of any figure (shape) is called the **perimeter** of that figure.

To find the perimeter of any figure, just add up the lengths of each of the sides of the figure.

Example   Find the perimeter of this figure.

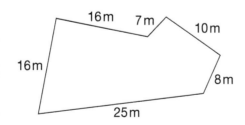

Perimeter = (16 + 25 + 8 + 10 + 7 + 16) m
          = 82 m

Sometimes, we are not given the measurements of all the sides, so we must first use what we do know to calculate the missing measurements.

Example   Find the perimeter of this figure.

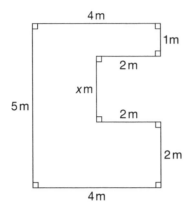

We calculate the missing length x first:

$x = (5 - 1 - 2)$

$ = 2$

Perimeter $= (5 + 4 + 2 + 2 + x\ [= 2] + 2 + 1 + 4)\,\text{m}$

$ = 22\,\text{m}$

E Perimeter of a rectangle, square, parallelogram and rhombus

Each **rectangle** (and **parallelogram**) has **two longer sides** and **two shorter sides**.

The two longer sides are equal in length, and the two shorter sides are equal in length.

The two longer sides are called the **length** of the rectangle (or parallelogram).

The two shorter sides are called the **width** of the rectangle (or parallelogram).

So the perimeter of a rectangle (or parallelogram)
$= 2 \times \text{length} + 2 \times \text{width}$

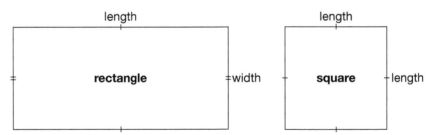

Each **square** (and **rhombus**) has **four sides** that are all exactly the **same length**.

So the perimeter of a square (or rhombus) $= 4 \times$ (length of one side)

Exercise 2

1 Find the perimeter of each of these figures. Remember to find any missing lengths first.

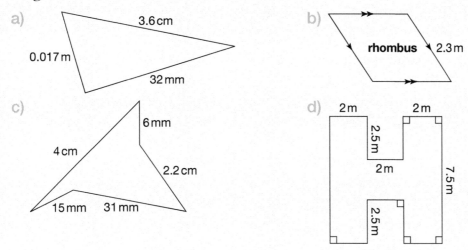

a)

3.6 cm

0.017 m

32 mm

b)

rhombus 2.3 m

c)

6 mm

4 cm

2.2 cm

15 mm   31 mm

d)

2 m   2 m

2.5 m

2 m

7.5 m

2.5 m

2 Find the length of the side of a square with a perimeter of 32 cm.

3 Find the width of a rectangle that has a perimeter of 24 cm and a length of 8 cm.

4 Find the length of a parallelogram that has a perimeter of 16.6 m and shorter side of length 3.1 m.

5 Find the perimeter of each of these figures. Remember to find any missing lengths first.

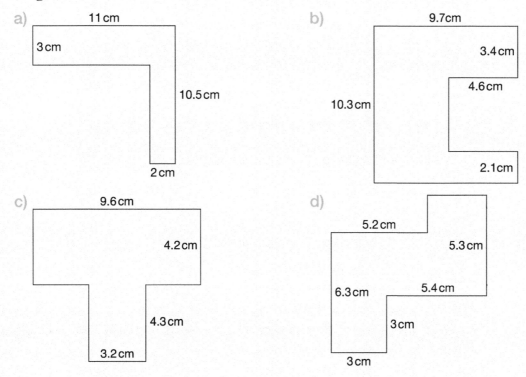

a)

11 cm

3 cm

10.5 cm

2 cm

b)

9.7 cm

3.4 cm

4.6 cm

10.3 cm

2.1 cm

c)

9.6 cm

4.2 cm

4.3 cm

3.2 cm

d)

5.2 cm

5.3 cm

6.3 cm   5.4 cm

3 cm

3 cm

F Area

Area is the word we use to describe the amount of space inside a plane figure.

We measure area using 'squares' that have a size we already know. In the metric system there are many units of area that we can use:

- 1 **square millimetre** ($1\,mm^2$) is the area of a square that has sides $1\,mm$ long.
  This is a very small area − about the size of a pinhead.
- 1 **square centimetre** ($1\,cm^2$) is the area of a square that has sides $1\,cm$ long.
  This is also a small area − about the size of your fingernail.
- 1 **square metre** ($1\,m^2$) is the area of a square that has sides $1\,m$ long.
  This is a much bigger area − about the size of four of your desks.
- 1 **square kilometre** ($1\,km^2$) is the area of a square that has sides $1\,km$ long.
  $1\,km = 1000\,m$
  So $1\,km^2 = 1000\,m \times 1000\,m = 1\,000\,000\,m^2$ (1 million square metres)
  This is a very big area − about the size of 200 football fields!
- 1 hectare (1 ha) is the area of a square that has sides $100\,m$ long.
  1 hectare $= 100\,m \times 100\,m = 10\,000\,m^2$
  The size of property (like farms) is often measured in hectares.

NOTE: In some countries, other units of area are used. For example, in Thailand, the size of farms is measured in 'rai'.
One rai is the area of a square that has sides $40\,m$ long.
1 rai $= 40\,m \times 40\,m = 1600\,m^2$

G Area of rectangles and squares

Look at the rectangle and the square below.

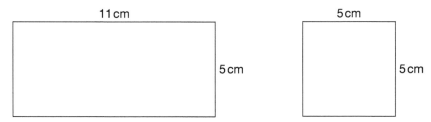

If we divide each of these figures into squares that are $1\,cm$ by $1\,cm$ (with area $1\,cm^2$) then we can calculate the area of the rectangle and the big square.

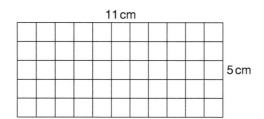 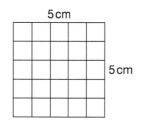

If we count the small squares inside the rectangle, we find there are 55. Each small square has an area of $1\,cm^2$, so the total area of the rectangle is $55\,cm^2$.

In the same way, we find that the total area of the big square is $25\,cm^2$.

In the rectangle there are 11 squares in each row, and there are 5 rows. To find the total number of small squares, we can multiply the number of squares in a row by the number of rows.

$$\text{Area of the rectangle} = (11 \times 5)\,cm^2$$
$$= 55\,cm^2$$

In the big square there are 5 squares in each row, and there are 5 rows. To find the total number of small squares, we can multiply the number of squares in a row by the number of rows.

$$\text{Area of the square} = (5 \times 5)\,cm^2$$
$$= 25\,cm^2$$

In both cases, the number of small squares in a row is the same as the length of the rectangle or square, and the number of rows is the same as the width of the rectangle (or the length of the square again).

So we can find the area of a rectangle by multiplying its length by its width.

$$\text{Area of a rectangle} = \text{length} \times \text{width}$$

NOTE: The length and width must be measured in the same units.

We can find the area of a square by multiplying its length by itself.

$$\text{Area of a square} = \text{length} \times \text{length} \quad \text{or} \quad (\text{length})^2$$

Some figures look quite complicated but can be 'divided' into a number of rectangles and/or squares. Then it is simple to work out the area of each of these rectangles and squares. The total area of the whole figure is then the **sum** of the areas of all the rectangles and/or squares.

Example

This is the shape of the lawn in Jack's garden. Calculate the area of the lawn in Jack's garden.

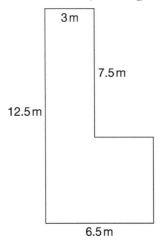

First, we must 'divide' the lawn into two rectangles, **A** and **B** (use a broken line ----------).

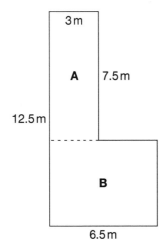

Area of rectangle **A** = $(3 \times 7.5)\,\text{m}^2$
             = $22.5\,\text{m}^2$

Area of rectangle **B** = $[(12.5 - 7.5) \times 6.5]\,\text{m}^2$
             = $(5 \times 6.5)\,\text{m}^2$
             = $32.5\,\text{m}^2$

Total area of lawn = $(22.5 + 32.5)\,\text{m}^2$
          = $55\,\text{m}^2$

Exercise 3

1 Calculate the missing numbers in each of these tables.

	Length of side	Area of square
a)	7 m	
b)	3.5 m	
c)		64 m²
d)		144 cm²

	Length	Width	Area of rectangle
e)	11 cm	13 cm	
f)	2.3 cm	1.4 m	
g)		7 cm	84 cm²
h)	8 m		48 m²
i)	5 cm		18 cm²
j)		1.2 m	2.88 m²

2 Calculate the area of each of these figures.

a)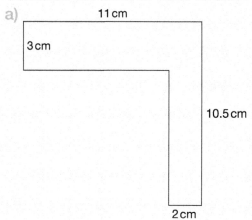

b)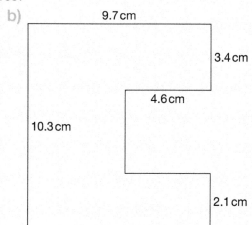

c)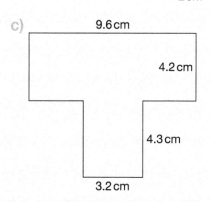

d)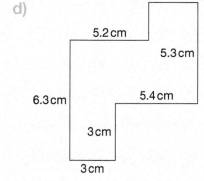

3 This is a plan of Phil's garden.
   a) Find the area of the whole garden.
   b) Find the area of the pond.
   c) Find the area of the paving.
   d) If the rest of the garden is grass, find the area of the grass.

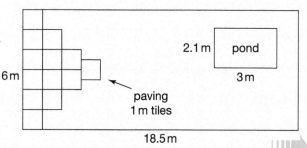

4 A square of cardboard has an area of 64 cm². A piece 1 cm wide is cut off all four sides.
What is the area of the new piece of cardboard?

5 This is a plan of a supermarket and its car park.

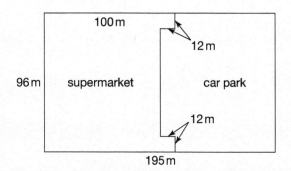

a) Find the area of the supermarket.
b) Find the area of the car park.
c) Each car needs 15 m² of space.
How many cars will fit in the car park?

6 This is a plan of Pippin Apple Farm and Busbey Woods.

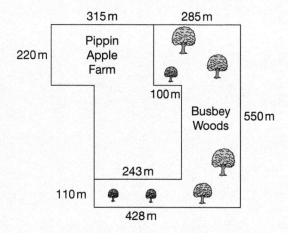

a) i) Find the area of Pippin Apple Farm.
   ii) Find the area of Busbey Woods.
b) The farmer now buys part of the woods and turns it into a field for his farm.
The new field measures 220 m by 100 m.
   i) What is the new area of Pippin Apple Farm?
  ii) What is the new area of Busbey Woods?

7 Here is a plan of part of Kidalots School.

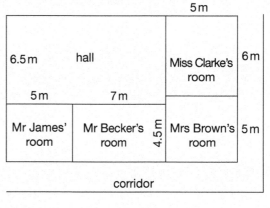

a) Find the area of Miss Clarke's room.
b) Find the area of Mrs Brown's room.
c) Find the area of Mr Becker's room.
d) Find the area of Mr James' room.
e) Find the area of the hall.
f) What is the area of the corridor if it is 2 m wide?
g) If carpet costs £20 per m², how much will it cost to carpet the corridor?

8 A school garden measures 10 m by 12 m. There is a path of 1 m wide that runs all the way round the garden just inside the boundary.

Calculate the area of the path.

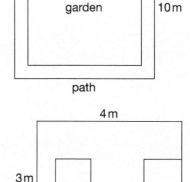

9 One wall of a room is 3 m high and 4 m wide. There is a door 1 m wide and 2 m high, and a window that is 1 m by 1 m.

a) Calculate the area of the wall to be painted.
b) 5 litres of paint will cover 30 m² of wall. Calculate how many litres of paint you will need to paint this wall.
c) The paint costs $5 per m². You paint the wall with two coats. How much will it cost?

10 A rectangular property is 120 m long and 70 m wide. On the property is a house that is a square of side 40 m. Calculate the area of the garden.

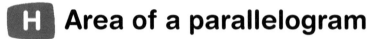

H Area of a parallelogram

We can find out how to work out the area of a **parallelogram** by comparing it to a **rectangle**.

Start with a parallelogram.

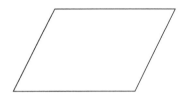

Draw a perpendicular line from one vertex to the opposite side.
This will make a triangle, **A**.

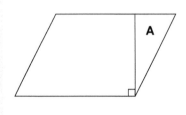

Now move triangle **A** to the opposite side of the parallelogram.
This makes a rectangle with the same area as the parallelogram.

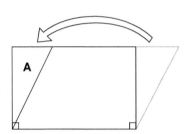

This 'new' rectangle has two sides that are the same length as two sides of the original parallelogram.
The other sides of the rectangle are equal to the perpendicular height (shortest distance) between the first two sides.

Area of a parallelogram = area of the rectangle
= base × height
where the base is the length of one pair of sides of the parallelogram, and the height is the perpendicular distance between these two sides.

NOTE: Remember that both lengths must be in the same units.

Example

Find the area of this parallelogram.

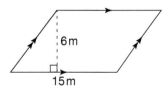

Area of a parallelogram = base × height
= 15 m × 6 m
= 90 m^2

Sometimes we are given the area and one of the measurements. We can work backwards to find the other measurement using one of these formulae:

$$\text{Height} = \frac{\text{area}}{\text{base}}$$

$$\text{Base} = \frac{\text{area}}{\text{height}}$$

Examples   A parallelogram has an area of $48\,\text{cm}^2$.

a) If the base is $8\,\text{cm}$, calculate the height.

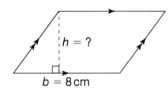

$$\text{Height} = \frac{\text{area}}{\text{base}}$$

$$= \frac{48}{8}$$

$$= 6$$

The height is $6\,\text{cm}$.

b) If the height is $10\,\text{cm}$, calculate the base.

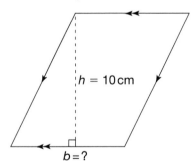

$$\text{Base} = \frac{\text{area}}{\text{height}}$$

$$= \frac{48}{10}$$

$$= 4.8$$

The base is $4.8\,\text{cm}$.

▌ Area of a triangle

We now know that the area of a parallelogram = base × height.
We can use this to help us work out the area of a **triangle**.

If we draw any triangle and then copy it exactly, we will find that these
two identical triangles together make a **parallelogram** (if the triangle is
a right-angled one, then the 'parallelogram' will actually be a rectangle).

Draw a triangle.   Copy this triangle and rotate   Join the two triangles to
                      the copy through 180° as   make a parallelogram.
                      shown by the arrow.

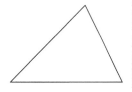

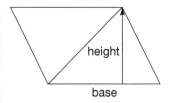

If two identical triangles make a parallelogram with the same base
and perpendicular height, then the area of one of the triangles will be
half the area of the parallelogram.

Area of a triangle = $\frac{1}{2}$ base × height

or $\dfrac{\text{base} \times \text{height}}{2}$

where the height is the perpendicular distance from the
base to the opposite vertex.

NOTE: Remember that both lengths must be in the same units.

Examples   Calculate the area of each of these triangles.

a)
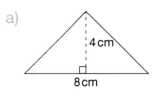

Area = $\frac{1}{2}$ × base × height
     = $\frac{1}{2}$ × 8 × 4
     = $\frac{1}{2}$ × 32 = 16
The area is 16 cm².

b)

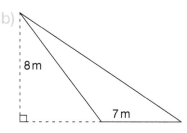

Area $= \frac{1}{2} \times$ base $\times$ height

$\quad\quad = \frac{1}{2} \times 7 \times 8$

$\quad\quad = \frac{1}{2} \times 56 = 28$

The area is $28\,\text{m}^2$.

Sometimes we are given the area and one of the measurements. We can work backwards to find the other measurement using one of these formulae:

$$\text{Height} = \frac{2 \times \text{area}}{\text{base}}$$

$$\text{Base} = \frac{2 \times \text{area}}{\text{height}}$$

Examples

A triangle has an area of $60\,\text{cm}^2$.

a) If the base is 10 cm, calculate the height.

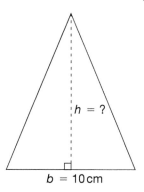

$$\text{Height} = \frac{2 \times \text{area}}{\text{base}}$$

$$= \frac{2 \times 60}{10} = 12$$

The height is 12 cm.

b) If the height is 6 cm, calculate the base.

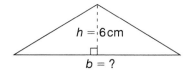

$$\text{Base} = \frac{2 \times \text{area}}{\text{height}}$$

$$= \frac{2 \times 60}{6} = 20$$

The base is 20 cm.

Exercise 4

1 Calculate the area of each of these shapes.

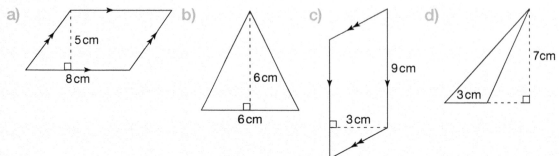

a) 5 cm, 8 cm

b) 6 cm, 6 cm

c) 9 cm, 3 cm

d) 7 cm, 3 cm

2 Calculate the area of each of these shapes.

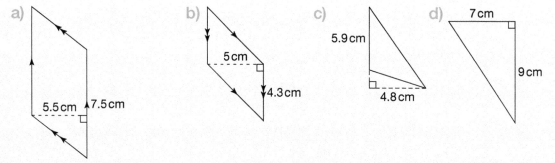

a) 5.5 cm, 7.5 cm

b) 5 cm, 4.3 cm

c) 5.9 cm, 4.8 cm

d) 7 cm, 9 cm

3 In each of these questions, two of the three figures have the same area. Calculate the areas of all the figures to find which ones are the same.

a)   i) Triangle with a base of 8 cm and a height of 6 cm
    ii) Parallelogram with a base of 6 cm and a height of 3 cm
    iii) Triangle with a base of 9 cm and a height of 4 cm

b)   i) Triangle with a base of 8 cm and a height of 5 cm
    ii) Parallelogram with a base of 7 cm and a height of 5 cm
    iii) Triangle with a base of 10 cm and a height of 4 cm

c)   i) Triangle with a base of 12.6 cm and a height of 6.3 cm
    ii) Triangle with a base of 13.23 cm and a height of 6 cm
    iii) Parallelogram with a base of 9.4 cm and a height of 7.2 cm

4 Copy and complete this table.

Base	Height	Area of triangle
	10 mm	40 mm²
16 cm		80 cm²
9 cm		27 cm²
	6 cm	21 cm²
9 mm		18 mm²
	7 mm	28 mm²

5 In this figure, $BK = 10\,cm$, $AC = 12\,cm$ and $BC = 15\,cm$.

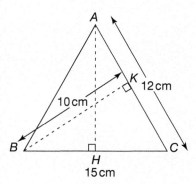

a) Find the area of triangle ABC.

b) Use your answer to part **a)** to calculate the height, AH.

6 $ABCD$ is a rectangle.

a) Find the area of triangle AEF.

b) Use this to calculate the area of $BCDFE$.

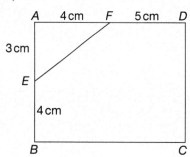

7 Copy and complete this table.

Base	Height	Area of parallelogram
7 cm		56 cm²
	5 m	45 m²
	8 cm	24 cm²
6 mm		60 mm²

8 $ABCD$ is a parallelogram with $AB = 6\,cm$, $BC = 8\,cm$ and $AH = 4\,cm$.

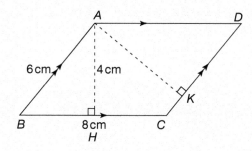

a) Calculate the area of the parallelogram.

b) Use this to find the height AK.

J Area of a trapezium

To find out how to work out the area of a trapezium that has parallel sides a and b, we must first cut this trapezium in half and re-arrange the parts.

The trapezium has parallel sides a and b. The perpendicular distance between these sides is equal to h.

Cut the trapezium in half across the middle.

We now have two new trapeziums. The height of each is half the height, h, of the trapezium we started with, or $\dfrac{h}{2}$.

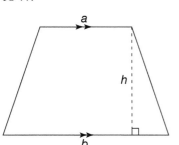

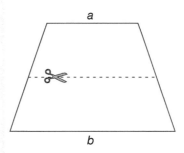

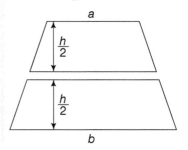

Rotate the top new trapezium through 180° as shown by the arrow. Join up the two trapeziums.

This makes a parallelogram with base $(a + b)$, and height $\dfrac{h}{2}$.

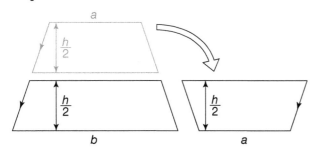

We know that the area of a parallelogram = base × height

So the area of this parallogram = $(a + b) \times \dfrac{h}{2}$

This is the same as taking the sum of the lengths of the parallel sides and multiplying it by the height, and dividing the result by 2.

$$\text{Area of a trapezium} = \tfrac{1}{2} \times (\text{sum of the lengths of the parallel sides}) \times \text{height}$$

$$= \tfrac{1}{2}(a + b)h \quad or \quad \dfrac{h(a + b)}{2}$$

where a and b are the lengths of the **parallel sides** and h is the **perpendicular height** between them.

NOTE: Remember that a, b and h must all be in the same units.

Example   Calculate the area of this trapezium.

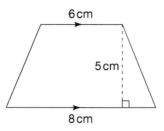

Area of trapezium $= \dfrac{h(a + b)}{2}$

$\qquad\qquad\qquad = \dfrac{5 \times (6 + 8)}{2} = \dfrac{70}{2} = 35$

The area of the trapezium is $35\,\text{cm}^2$.

Sometimes we are given the area and two of the measurements. We can work backwards to find the other measurement.

Example   A trapezium has an area of $60\,\text{cm}^2$.
It has a height of $6\,\text{cm}$, and the longer of the two parallel sides measures $12\,\text{cm}$.

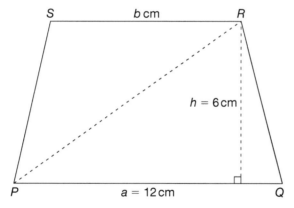

Calculate the length of the other parallel side.

Area of $\triangle PQR = \frac{1}{2} \times 12 \times 6 = 36$

The area of $\triangle PQR$ is $36\,\text{cm}^2$.

Area of $\triangle PRS$ = area of trapezium $-$ area of $\triangle PQR$

$\qquad\qquad\quad = 60 - 36 = 24$

The area of $\triangle PRS$ is $24\,\text{cm}^2$.

Base of $\triangle PRS = \dfrac{2 \times \text{area of } \triangle PRS}{\text{height}}$

$\qquad\qquad\quad = \dfrac{2 \times 24}{6} = 8$

The base of $\triangle PRS$ is $8\,\text{cm}$.

K Working out the area of compound figures

In real life, we sometimes need to work out the area of a simple figure (e.g. rectangle, square, triangle, parallelogram, rhombus, trapezium). Often, however, the shape is made up of several of these simple figures joined together. A shape like this is called a compound figure.

To work out the **area** of a compound figure, follow these steps.

1 Make sure you have a clear drawing of the compound figure with all the lengths you know written on it.
2 Draw as many extra straight lines as you need so that the whole shape is divided into the simple figures that you know.
3 Work out the lengths of any lines that you do not already know.
4 Use the formulae you have learned to work out the area for each of the simple figures.
5 Add up the areas for all of these smaller pieces to find the total area of the compound figure.

NOTE: Sometimes it is easier to find the area of one simple figure and *subtract* it from the area of another bigger simple figure, to work out the area of your compound figure. For example, to find the area of this black compound figure, it is easier to subtract the area of the white triangle from the area of the whole black rectangle.

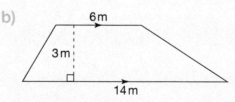
Exercise 5

1 Find the area of each of these trapeziums.

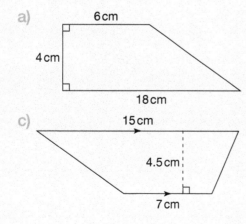

a) 6 cm
4 cm
18 cm

b) 6 m
3 m
14 m

c) 15 cm
4.5 cm
7 cm

2 A trapezium has an area of 60 cm². Calculate the height if the sum of the parallel sides is a) 10 cm b) 15 cm.

3 A trapezium has an area of 80 cm². Calculate the length of the second parallel side if:

a) it has a height of 8 cm and the shorter parallel side is 6 cm long

b) it has a height of 10 cm and the longer parallel side is 12 cm long.

4 Find the area of each of these compound figures.

a)
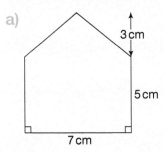
3 cm

5 cm

7 cm

b)
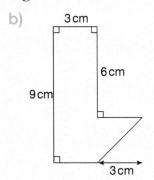
3 cm

6 cm

9 cm

3 cm

c)
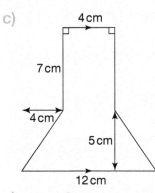
4 cm

7 cm

4 cm

5 cm

12 cm

d)
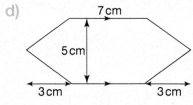
7 cm

5 cm

3 cm   3 cm

e)
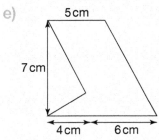
5 cm

7 cm

4 cm   6 cm

f)
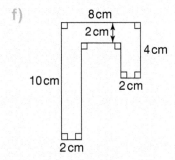
8 cm

2 cm

4 cm

10 cm

2 cm

2 cm

g)
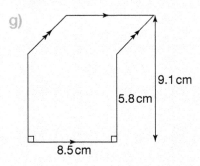
9.1 cm

5.8 cm

8.5 cm

h)
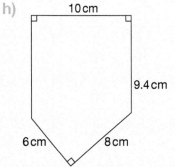
10 cm

9.4 cm

6 cm   8 cm

5 This is part of a stained glass window.

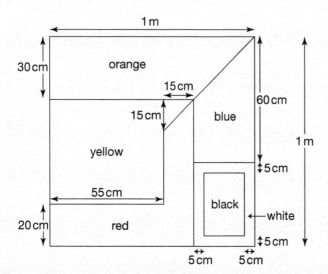

Work out the area of each piece of different coloured glass. When you have finished, add up the areas of all the coloured pieces to check that they equal the area of the whole square.

6 Here is a diagram of a compound figure.

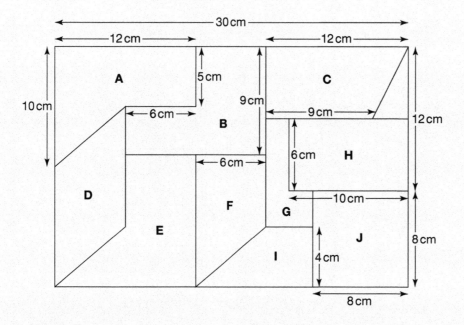

a) Work out the lengths of the horizontal and vertical lines that are not given and write these lengths on a copy of the diagram.

b) Work out the area of each of the pieces.

c) Add up the areas of all the pieces and check that they equal the total area of the rectangle.

 Circles

What do you remember about circles?
What should the missing words be?

A circle has _____ side (or maybe so
many that we cannot count them!).
Every point on a circle is the _____ from
the point called the centre.
This distance from the centre to the
outside of the circle is called the _____.
A line segment joining one point on a circle to another point on the
circle and passing through the centre is called the _____.
The length of the diameter is _____.
The length around the outside (perimeter) of the circle is called
the _____.
One part of the outside of the circle is called an _____.
There are _____° in a circle.

Look back at page 7 to check your answers.

Activity

Use some string and/or tape to measure the **circumference** and the **diameter** of each of
these items.
Then calculate circumference ÷ diameter for each item.

Item	Circumference (c)	Diameter (d)	$\frac{c}{d}$
Bicycle wheel			
Mug or glass			
Coin			

Your results for all the values of $\frac{c}{d}$ should be close to 3. If we take the average of all the results
in the class, we should find it is close to 3.14.

We can therefore say that, for any circle, $\frac{c}{d}$ is always equal to the same number (called a **constant**).

This is a very important constant, and it is given a special name and symbol.
We use the symbol π, which is the Greek letter **pi** (pronounced 'pie').
It is not possible to write an **accurate** number for the constant π. For our calculations we will
use two **approximate** values: $\pi \approx \frac{22}{7}$ and $\pi \approx 3.14$.

Circumference of a circle

We know that $\frac{c}{d} = \pi$.

So we can say that $c = \pi d$

$\qquad\qquad$ or $c = 2\pi r$ (because $d = 2r$).

$\qquad$ The circumference of a circle $= 2\pi r$
$\qquad$ where r is the radius of the circle.

Example
If the radius of a circle is 5 cm, calculate the circumference (use $\pi = 3.14$).

$c = 2\pi r$
$\quad = 2 \times 3.14 \times 5 = 31.4$
The circumference is 31.4 cm.

If we are given the circumference, we can find the radius (or diameter) by working backwards using this formula:

$$r = \frac{c}{2\pi}$$

Example
If the circumference of a circle is 22 cm, calculate the radius $\left(\text{use } \pi = \frac{22}{7}\right)$.

$r = \dfrac{c}{2\pi}$

$\quad = \dfrac{22}{2 \times \frac{22}{7}}$

$\quad = 22 \div \left(\dfrac{2 \times 22}{7}\right)$

$\quad = 22 \times \dfrac{7}{44}$

$\quad = \dfrac{7}{2} = 3.5$

The radius is 3.5 cm.

Sometimes we need to find the perimeter of a **semicircle** (= half a circle). It is important to remember to include the straight side!

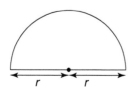

Example   Find the perimeter of a semicircle with radius $7\,cm$ $\left(use\ \pi = \frac{22}{7}\right)$.

$$Perimeter = \tfrac{1}{2}c + 2r$$
$$= \left(\tfrac{1}{2} \times 2\pi r\right) + 2r$$
$$= \tfrac{22}{7} \times 7 + (2 \times 7)$$
$$= 22 + 14 = 36$$

The perimeter is $36\,cm$.

Exercise 6

1 Use $\pi = \frac{22}{7}$ to calculate the circumference of each of these circles.

   a) Radius of $7\,cm$ b) Radius of $14\,cm$
   c) Radius of $3.5\,cm$ d) Radius of $4.9\,cm$

2 Use $\pi = \frac{22}{7}$ to find the radius of circles with these circumferences.

   a) $44\,cm$ b) $132\,cm$
   c) $8.8\,cm$ d) $2.2\,cm$

3 Use $\pi = \frac{22}{7}$ to calculate the perimeter of each of these figures.

   a)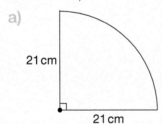

   b)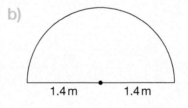

   c)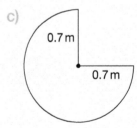

4 Use $\pi = 3.14$ to calculate the perimeter of each of these figures (O is the centre of the semicircle).

   a)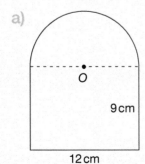

   b)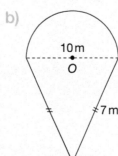

Area of a circle

It is quite difficult to show exactly how to work out the area of a circle. The ancient Greek mathematicians worked on the problem for a long time. They found that:

Area of a circle $= \pi r^2$
where r is the radius of the circle.

Example

Use $\pi = \frac{22}{7}$ to calculate the area of a circle with a radius of 7 cm.

Area of circle $= \pi r^2$
$$= \frac{22}{7} \times 7 \times 7 = 154$$

The area of the circle is 154 cm².

If we are given the area, we can find the radius (or diameter) by working backwards by using this formula:

$$r^2 = \frac{\text{Area}}{\pi}$$

Example

Use $\pi = \frac{22}{7}$ to calculate the radius of a circle with an area of 616 cm².

$$r^2 = \frac{\text{Area}}{\pi}$$
$$= \frac{616}{\frac{22}{7}}$$
$$= 616 \div \frac{22}{7}$$
$$= 616 \times \frac{7}{22}$$
$$= 196$$

So $r = \sqrt{196}$ cm
$$= 14 \text{ cm}$$

We can also find the area of compound figures made using circles and semicircles.

Examples

a) Use $\pi = 3.14$ to calculate the area of the figure if O is the centre of the semicircle.

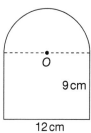

9 cm

12 cm

Area of figure = area of semicircle + area of rectangle
$$= \left(\tfrac{1}{2} \times \pi r^2\right) + (12 \times 9)$$
$$= \left(\tfrac{1}{2} \times 3.14 \times 6 \times 6\right) + 108$$
$$= 56.52 + 108 = 164.52$$

The area of the figure is $164.52\,\text{cm}^2$.

b) Use $\pi = 3.14$ to calculate the area of the shaded part of the figure.

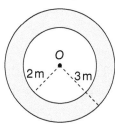

Shaded area = area of big circle − area of small circle
$$= \pi(3)^2 - \pi(2)^2$$
$$= 9\pi - 4\pi$$
$$= 5\pi$$
$$= 5 \times 3.14 = 15.7$$

The shaded area is $15.7\,\text{cm}^2$.

Exercise 7

1 Use $\pi = \frac{22}{7}$ to calculate the area of each of these circular objects.

   a) A circular saucer of radius 7 cm   b) A circular coin of radius 1.4 cm
   c) A circular plate with a diameter of 21 cm   d) A circle with a diameter of 7 cm

2 Use $\pi = \frac{22}{7}$ to calculate the radius of each of these circular objects.

   a) A circular plate with an area of $154\,\text{cm}^2$

   b) A circular pond with an area of $38\tfrac{1}{2}\,\text{m}^2$

3 Find the area of the shaded part of each of these compound figures.

   a)

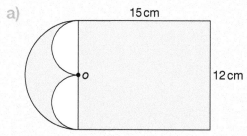

Use $\pi = 3.14$.
Point O is the centre of the shaded circle.
Each white piece is a semicircle.

b)

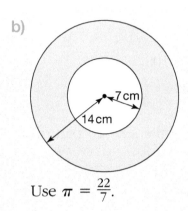

Use $\pi = \frac{22}{7}$.

c)

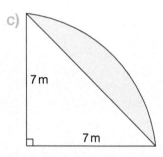

Use $\pi = \frac{22}{7}$.
The right-angled vertex is the centre of the shaded circle.

4 Mr Jones has a rectangle of lawn in his garden. The lawn measures 15 m by 10 m.
He wants to build a circular swimming pool with a diameter of 4 m in the middle of the lawn.
What area of lawn will he have left?
Use $\pi = 3.14$.

5 You have a circular metal disk with a radius of 12 cm.
You cut out a circular disk of radius 6 cm. What is the area of the disk left over?
Use $\pi = 3.14$.

6 Calculate the area of the shaded part of this figure.

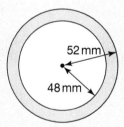

Use $\pi = 3.14$.

7 This is a semicircle with a radius of 3 cm.

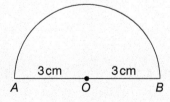

a) Calculate the area of this semicircle
b) Calculate the perimeter of this semicircle.
    Use $\pi = 3.14$.

Unit 2 — Algebraic equations in one variable

Key vocabulary

backwards	expression	solve
balanced	inspection	variable
equation		

A What is an equation?

Can you solve these puzzles?

- Zoinck is a number they use on Mars!
  If you add 3 to zoinck you get 12.
  Or 3 + zoinck = 12
  What is zoinck in our earth numbers?
- What number must we put in each box to make these statements correct?

  $\boxed{} + 3 = 8$　　$\boxed{} \times 3 = 30$　　$2 \times \boxed{} - 3 = 7$

These are all examples of equations.

An equation is basically a puzzle that has to be solved. It has a **left-hand side (LHS)** and a **right-hand side (RHS)**, separated by an **equals sign**.

B Solving equations by inspection

Equations like the ones above are so simple that we can usually find the answer just by looking at them and working out it in our head. This method is known as inspection.

Instead of words or boxes, equations are usually written using **letters** for the unknown numbers.
We call these letters variables. These are the same kind of variables that we used to write algebraic expressions.

So equations can have an algebraic expression on one side, or sometimes on both sides. These kinds of equations are called **algebraic equations**.

To solve an equation we must find the numerical value of the letter (variable) that fits the equation to make it **true**.

Examples

Solve these equations by inspection.

a) $x - 2 = 6$

So $x = 8$

Check: $\underline{8} - 2 = 6$ ✓

b) $2y = 10$

So $y = 5$

Check: $2 \times \underline{5} = 10$ ✓

c) $6(n + 1) = 60$

So $n = 9$

Check: $6(\underline{9} + 1) = 60$ ✓

Exercise 1

1 Copy these statements. Write a number in each box to make the statements true.

a) $\boxed{} + 4 = 7$

b) $15 - \boxed{} = 11$

c) $13 = \boxed{} + 4$

d) $11 = \boxed{} - 5$

e) $3 \times \boxed{} = 15$

f) $24 = 8 \times \boxed{}$

g) $\boxed{} \div 2 = 9$

h) $7 = \boxed{} \div 3$

2 Copy these statements. Write a number in each box to make the statements true.

a) $3 \times \boxed{} = 18$

b) $24 = \boxed{} \times 4$

c) $\boxed{} \div 2 = 7$

d) $2 \times \boxed{} + 6 = 12$

e) $15 - 3 \times \boxed{} = 9$

f) $29 = 4 \times \boxed{} - 3$

g) $\dfrac{\boxed{}}{2} + 3 = 7$

h) $\dfrac{\boxed{}}{3} - 2 = 3$

i) $7 - \dfrac{8}{\boxed{}} = 5$

3 Solve these equations by inspection.

a) $a + 7 = 10$

b) $y - 4 = 4$

c) $4c = 20$

d) $8 = \dfrac{d}{2}$

e) $x - 2 = 6$

f) $11 - b = 4$

g) $2y = 10$

h) $7 = \dfrac{x}{4}$

i) $3x + 1 = 10$

j) $4d - 7 = 13$

k) $31 = 6y + 7$

l) $2(c + 3) = 24$

m) $12 = 3(5 - a)$

n) $3y + 2 = 8$

o) $2d - 4 = 6$

p) $8a + 7 = 79$

C Solving equations by working backwards

I think of a number, multiply it by 3 and add 4.
The answer is 19.
What is the number I thought of?

Imagine that x is the number I thought of.
The steps of the problem can be shown in a diagram:

Remember:

Forwards	Backwards
add	subtract
subtract	add
multiply	divide
divide	multiply

$x \longrightarrow$ [multiply by 3] $\xrightarrow{} 3x \xrightarrow{}$ [add 4] $\xrightarrow{} 3x + 4 \xrightarrow{}$ Answer 19

Now, work backwards, doing the opposite calculation each time:

$5 \longleftarrow$ [divide by 3] $\xleftarrow{} 15 \xleftarrow{}$ [subtract 4] $\xleftarrow{} 19$

I thought of the number 5.

This is an approximate rule for changing temperatures in degrees Celsius to temperatures in degrees Fahrenheit:

Fahrenheit = double Celsius and add on 30

Examples

a) Find the value of F when $C = 6$.

When $C = 6$
double $C = 2 \times 6 = 12$
and add on 30: $12 + 30 = 42$
So when $C = 6$, $F = 42$.

b) What is the value of C when $F = 58$?

To find C we must use the rule in reverse.

$C \longrightarrow$ [double] $\xrightarrow{} 2C \xrightarrow{}$ [add on 30] $\xrightarrow{} 2C + 30 \xrightarrow{}$ Answer 58

$14 \longleftarrow$ [divide by 2] $\xleftarrow{} 28 \xleftarrow{}$ [subtract 30] $\xleftarrow{} 58$

So when $F = 58$, $C = 14$.

c) Write down a formula for F in terms of C.

$F = 2C + 30$

1 I think of a number and then double it. The answer is 10.
What is my number?

2 I think of a number, double it and add 4. The answer is 16.
What is my number?

3 I think of a number, multiply it by 5 and then add 2. The answer is 17.
What is my number?

4 I think of a number, add 4 then double the result. The answer is 24.
What is my number?

5 I think of a number, multiply it by 3 and then subtract 5. The answer is 7.
What is my number?

6 I think of a number, double it and add 3, then multiply the result by 4.
The answer is 52.
What is my number?

7 Kathryn thinks of a number. She adds 3 and then doubles the answer.

   a) What number does Kathryn start with to get an answer of 10?
   b) What is her answer if she starts with x?

8 Sarah thinks of a number x. She subtracts 2 from it and multiplies the result
by 3.
What is her answer?

9 Lauren uses this rule:

   Start with a number. Add 2. Multiply by 3. Write down the result.

   a) What is the result when Lauren starts with 5?
   b) What is the result when Lauren starts with -5?
   c) What is the result when Lauren starts with x?

10 An approximate rule for changing kilometres, K, into miles, M, is:

   Multiply the number of kilometres by 5, then divide by 8 to give
   the number of miles.

   a) Find the value of M when
      i) $K = 24$
      ii) $K = 60$
      iii) $K = 18$
   b) What is the value of K when M = 10?
   c) What is the value of K when M = 32?
   d) Write down the equation for M in terms of K.

11 The rule to find the cooking time, C minutes, of a chicken that weighs k kg is:

Multiply the mass of the chicken (in kg) by 40 and then add 20.

a) Find the cooking time for a chicken that weighs 3 kg.
b) Find the weight of a chicken that has a cooking time of 100 minutes.
c) Write an equation for C in terms of k.

12 The cost of a taxi journey is calculated like this:

£3 plus £2 for each kilometre travelled.

a) Alex travels 5 km by taxi. How much does it cost?
b) A taxi journey of k km costs £C. Write an equation for cost, C, in terms of k.

13 The cost of hiring a ladder is given by this rule:

£12 per day, plus a delivery charge of £8.

a) Ahmed hired a ladder for three days. How much did he pay?
b) Simone hired a ladder for six days. How much did he pay?
c) Jin hired a ladder for x days. Write down an equation for the total cost, C, in terms of x.

 ## Solving equations using the 'balance method'

It is not always possible to solve equations just by inspection or working backwards. Many equations are more difficult and we need a system to help solve them.

Remember, what we are trying to do when we solve an equation is to find the numerical value of a variable (letter), by ending up with **one letter** on one side of the equation and a **number** on the other side of the equation.

Because the LHS and RHS are equal (it is an **equation**), we can also say the two sides are balanced.

If we perform the **same** mathematical operation to **both** sides of the equation, it will still be **balanced**. This means we can:

● **add** the same number or term to both sides,
● **subtract** the same number or term from both sides,
● **multiply** or **divide** both sides by the same number, and the equation will remain balanced.

Sometimes, an equation includes **brackets**. Before using the balance method, any brackets must be simplified by multiplying out (called **expanding the brackets**). Once the brackets have been removed, the balance method can be used as normal.

If the equation contains any **fractions**, remove the fractions **first** by multiplying both sides by the **lowest common multiple of the denominators** of all the fractions.

Examples

Solve these equations.

a) $4(3 + 2x) = 5(x + 2)$

$\quad 12 + 8x = 5x + 10$ First expand the brackets.

$\quad\quad\quad 8x = 5x - 2$ Subtract 12 from both sides.

$\quad\quad\quad 3x = -2$ Subtract $5x$ from both sides.

$\quad\quad\quad x = -\frac{2}{3}$ Divide both sides by 3.

b) $\quad\quad \dfrac{x}{2} + \dfrac{2x}{3} = 7$

$6 \times \dfrac{x}{2} + 6 \times \dfrac{2x}{3} = 6 \times 7$ The LCM of all the denominators is 6, so multiply both sides by 6.

$\quad\quad 3x + 4x = 42$

$\quad\quad\quad 7x = 42$ Combine like terms.

$\quad\quad\quad x = 6$ Divide both sides by 7.

c) $\quad\quad \dfrac{x - 1}{3} = \dfrac{x + 1}{4} - \dfrac{3(x - 2)}{2}$

$12 \times \left(\dfrac{x - 1}{3} \right) = 12 \times \left[\dfrac{x + 1}{4} - \dfrac{3(x - 2)}{2} \right]$ The LCM of all the denominators is 12, so multiply both sides by 12.

$4(x - 1) = 3(x + 1) - 18(x - 2)$

$4x - 4 = 3x + 3 - 18x + 36$ Expand each set of brackets.

$4x - 4 = 39 - 15x$ Combine like terms.

$4x = 43 - 15x$ Add 4 to both sides.

$19x = 43$ Add $15x$ to both sides.

$x = \dfrac{43}{19}$ Divide both sides by 19.

$\quad = 2\frac{5}{19}$

Exercise 3

1 Solve each of these equations.

a) $3(n + 5) + n = 23$

b) $3(2z - 5) = z + 15$

c) $4(2w + 3) + 7 = 43$

d) $m + 2(m + 1) = 14$

e) $2(3b - 4) = 3(b + 1) - 5$

f) $2(3 - 2x) = 2(6 - x)$

g) $2(3w - 1) + 4w = 28$

h) $2(y + 4) + 3(2y - 5) = 5$

i) $3(2v + 3) = 5 - 4(3 - v)$

j) $5c - 2(4c - 9) = 5 + 5(2 - c)$

k) $5(x + 2) + 2(2x - 1) = 7(x - 4)$

l) $3(x - 4) = 5(2x - 3) - 2(3x - 5)$

2 Solve each of these equations.

a) $\dfrac{4 - d}{3} = \dfrac{1}{-2}$

b) $\dfrac{2 - 3b}{3} = -\dfrac{5}{6}$

c) $\dfrac{2(4x - 3)}{5} = -6$

d) $\dfrac{x + 2}{5} = \dfrac{3 - x}{4}$

e) $\dfrac{2x}{3} - \dfrac{x}{6} = -2$

f) $\dfrac{x + 1}{2} + \dfrac{x - 1}{3} = 1$

g) $\dfrac{x + 2}{3} - \dfrac{x + 1}{4} = 2$

h) $\dfrac{11 - x}{4} = 2 - x$

i) $\dfrac{x + 2}{2} + \dfrac{x - 1}{5} = \dfrac{1}{10}$

j) $\dfrac{2x - 3}{6} + \dfrac{x + 2}{3} = \dfrac{5}{2}$

E Writing equations to solve word problems

So far you have been given the equations to solve.

The next step is to learn to write an equation using the information given in a problem.

The equations can then be solved in the usual way.

You may also be asked to use the solution to the equation to answer more questions about the same problem.

These problems must be solved in a very organised way.

Here are some steps that will help you to solve most of the problems.

1 Work in an **organised** manner. Read the **whole problem** first.

2 Analyse the problem. Write down what is **given**, and what must be **found**.

3 Draw a diagram if appropriate − it can really help to 'see' the problem.

4 Choose a **variable** (e.g. x) to stand in place of what must be found, or the value connected with what must be found.

5 Change the **word sentence** in the problem into a **maths sentence**. Use symbols so that you can write an **equation**.

6 **Solve** the equation to find the value of the unknown variable.

7 Translate this result back into **words** to answer the question asked.

8 **Check** your answer.

Examples

a) The triangle has sides of length x cm, $(x + 1)$ cm and $(2x − 3)$ cm.

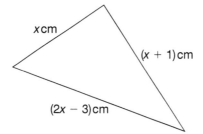

i) Write an expression, in terms of x, for the perimeter of the triangle.

Give your answer in its simplest form.

The perimeter of the triangle is $x + (x + 1) + (2x − 3)$ cm
$$= (4x − 2)\, \text{cm}$$

ii) The triangle has a perimeter of 18 cm.
Use an equation to find the value of x.

The perimeter of the triangle is 18 cm, so
$$4x − 2 = 18$$
$$4x = 20$$
$$x = 5$$

iii) Use this value of x to find the length of each side of the triangle.

$x = 5$
so $(2x − 3) = (2 \times 5 − 3)$
$$= 10 − 3$$
$$= 7$$
and $(x + 1) = 5 + 1$
$$= 6$$
So the lengths of the sides of the triangle are 5 cm, 7 cm and 6 cm.

b) **i)** Marco has a birthday party for 12 people.
How much does it cost?

> **Happy Days!**
> We will arrange your
> birthday party
>
> **Cost:** £20, plus £5 per person

$£20 + £5 \times 12 = £80$

ii) Toni has a birthday party for x people.
Write an equation for the cost C, in terms of x.

Cost $= £20 + £5 \times x$
$C = 20 + 5x$

iii) Jean pays £100 for her birthday party.
How many people went to the party?

$C = 20 + 5x,$
$100 = 20 + 5x$
$5x = 80$
$x = 16$

So 16 people went to Jean's birthday party.

c) Mabel has a balance scale, but only has a 5 g weight and a 50 g weight.

She puts one weight on each side of the scales, then keeps putting pencils on each side until the scales balance. She finds that seven pencils plus the 5 g weight balance with four pencils plus the 50 g weight.

How much does a pencil weigh?

Choose a letter to stand for the weight of one pencil in grams, say w.
Write an equation that is true for the information given:
$7w + 5 = 4w + 50$
Now solve the equation.

$7w + 5 = 4w + 50$
$7w = 4w + 50 - 5$
$7w = 4w + 45$
$7w - 4w = 45$
$3w = 45$
$w = 15$

So one pencil weighs 15 g.
Check: Seven pencils weigh $7 \times 15 = 105$ g, plus the 5 g weight, gives 110 g.
Four pencils weigh $4 \times 15 = 60$ g, plus the 50 g weight, gives 110 g.
So 15 g is correct.

d) Three-quarters of Keon's salary is $500 less than Naeem's salary.
If Naeem's salary is $2600, find Keon's salary.
Let Keon's salary be k.
So three-quarters of Keon's salary will be $\frac{3}{4}k$.
We are told that this is $500 less than Naeem's salary.
So we can say that Naeem's salary is $\frac{3}{4}k + 500$.
We know that Naeem's salary is $2600, so now we can write and solve an equation.

$2600 = \frac{3}{4}k + 500$

$2100 = \frac{3}{4}k$

$8400 = 3k$

$k = 2800$

So Keon's salary is $2800.
Check: $500 less than Naeem's salary is $2100.
$\frac{3}{4}$ of Keon's salary is $\frac{3}{4} \times \$2800 = \$2100$.
So $2800 is correct.

e) A field has an area of $630\,\text{m}^2$. The farmer wants to divide the field into three parts so that he can grow vegetables, flowers, and garden plants.
The area needed to grow garden plants is twice the area needed for flowers.
The area needed to grow vegetables is equal to the total area used for flowers and garden plants together.
Find the area needed to grow each type of plant.

Let the area needed to grow flowers be $x\,\text{m}^2$.
This means that the area needed to grow garden plants is $2x\,\text{m}^2$ and the area needed to grow vegetables is $(x + 2x)\,\text{m}^2$.
The total area of the field is $630\,\text{m}^2$, so

$x + 2x + (x + 2x) = 630$

$6x = 630$

$x = 105$

So the area needed to grow flowers is $105\,\text{m}^2$.
The area needed to grow garden plants is $(2 \times 105)\,\text{m}^2 = 210\,\text{m}^2$.
The area needed to grow vegetables is $(105 + 210)\,\text{m}^2 = 315\,\text{m}^2$.
Check: The total area used is $(105 + 210 + 315)\,\text{m}^2 = 630\,\text{m}^2$, as given.
So our calculations are correct.

f) Five-sixths of the difference between A's and B's test scores is equal to the sum of B's and C's scores. If B's test score is 40 points and C's score is 20 points, what is A's score?

Let A's test score be x points.
The difference between A's and B's scores is $x - 40$, and five-sixths of this is $\frac{5}{6}(x - 40)$.
The sum of B's and C's scores is $40 + 20 = 60$.

so $\frac{5}{6}(x - 40) = 60$
$5(x - 40) = 360$
$5x - 200 = 360$
$5x = 560$
$x = 112$

So A's test score is 112 points.

Check: $\frac{5}{6}(112 - 40) = \frac{5}{6} \times 72 = 60$

So our answer is correct.

Exercise 4

1 a) Write an expression, in terms of y, for the perimeter of this shape.

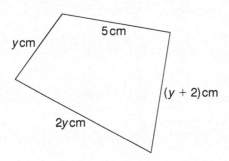

Give your answer in its simplest form.

b) The shape has a perimeter of 39 cm.
Write an equation to find the value of y.

2 a) Write an expression, in terms of x, for the sum of the angles of this triangle.

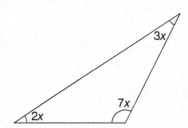

Give your answer in its simplest form.

b) The sum of the angles is 180°.
Write an equation to find the value of x.

3 The length of a rectangle is x cm. The width of the rectangle is 4 cm less than the length.

  a) Write an expression, in terms of x, for the width of the rectangle.
  b) Write an expression for the perimeter of the rectangle in terms of x.
  c) The perimeter of the rectangle is 20 cm. Write an equation to find the value of x.

4 A teacher uses this rule to work out the number of exercise books he needs for his pupils:

  3 books per pupil, plus 50 extra books.

  a) This year, he has 120 pupils. How many books does he need?
  b) Using n for the number of books and s for the number of pupils, write down the teacher's rule for n in terms of s.
  c) For next year, he will need 470 books. How many pupils will there be?

5 The cost of a pencil is x pence. The cost of a pen is 10 pence more than a pencil.

  a) Write an expression, in terms of x, for the cost of a pen.
  b) Write an expression, in terms of x, for the total cost of a pencil and two pens.
  Give your answer in its simplest form.
  c) The total cost of a pencil and two pens is 65 pence.
  Write an equation and use it to find the value of x.
  d) How much does a pencil cost?
  e) How much does a pen cost?

6 a) Write an expression, in terms of x, for the perimeter of this shape.

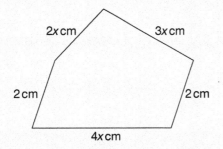

  b) The perimeter is 58 cm. Find the value of x.
  c) What is the length of the longest side of the shape?

7 Geoffrey knows that the sum of the angles in this shape adds up to 540°.

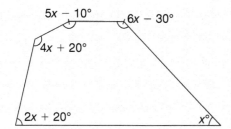

a) Write an equation to find the value of x.

b) Use your equation to find the size of the biggest angle.

8 a) Write an expression, in terms of x, for the area of the rectangle.

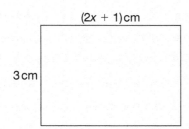

b) The rectangle has an area of $30\,cm^2$.
Write an equation to find the value of x.

c) What is the length of this rectangle?

9 a) Write an expression, in terms of t, for the perimeter of triangle **A**.
Give your answer in its simplest form.

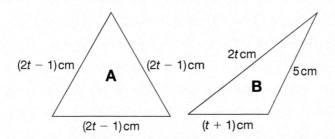

b) Write an expression, in terms of t, for the perimeter of triangle **B**.
Give your answer in its simplest form.

c) The perimeters of both triangles are equal.
Find the value of t.

d) Calculate the length of a side of triangle **A**.

10 x is an even number.
   a) What is the next even number after x?
   b) What is the even number before x?
   c) The sum of all three of these even numbers is equal to 48.
      Write an equation to find the value of each of the three numbers.

11 A woman bought some oranges to give to 15 people.
   Everybody did not receive the same number of oranges, so she bought 10 more oranges.
   Now, each person has four oranges.
   Find the number of oranges the woman bought the first time.

12 Dad added some extra money to Mom's $50, and shared the total among their five children.
   Each child received $130.
   Calculate how much extra money Dad added.

13 The difference between x and y is 10 more than the difference between y and z.
   The sum of x, y and z is 114 and $y = 4$.
   Calculate the value of x and z.

14 A square has sides of $\frac{2}{3}$ m. A rectangle has a width of $1\frac{1}{3}$ m and a length of $2\frac{1}{3}$ m.
   The area of the rectangle is equal to the area of the square plus the area of a triangle.
   Calculate the area of the triangle.

15 Pong has some goats. Half of his goats are mothers. Two-fifths of the other half of the goats are babies. The rest of the goats are adults with no babies.
   There are nine more mothers than babies.
   Calculate how many goats Pong has.

16 A mother is now five times as old as her daughter.
   Six years ago, the sum of their ages was 60.
   Find their ages now.

17 A boy is 12 years old. His father is four times as old.
   In x years' time, the father will be three times as old as the son.
   Calculate the value of x.

18 The sum of three consecutive odd numbers is 99. What are the numbers?

19 The cost of 3 kg of chicken and 4 kg of beef is $29.
The cost of 1 kg of chicken is $2 less than the cost of 1 kg of beef.
Find the cost of both the meats.

20 The cost of seven buns and eight slices of cake is $5.90.
The cost of a slice of cake is 20 ¢ less than the cost of a bun. Find the cost of a bun.

21 At a fruit stall, Mr Wu bought 1.4 kg of pears and Mr Li bought 1.9 kg of the same pears.
Mr Li paid $2.50 more than Mr Wu. Calculate the cost of 1 kg of pears at this stall.

22 Mrs Rajah bought the same weight (in kilograms) of both rice and sugar.
Rice costs $1.50 per kg, and sugar costs $0.90 per kg. Mrs Rajah paid $12.00 altogether.
Calculate how much (in kilograms) of each she bought.

23 The length of rectangle A is 3 times its width.
The length of rectangle B is 4 m shorter than the length of rectangle A.
The width of rectangle B is 1 m longer than the width of rectangle A.
The perimeter of rectangle B is 66 m.
Calculate the length and width of rectangle A.

24 The perimeter of a triangle is 69 cm.
Side A is 5 cm shorter than side B.
Side C is twice as long as side A.
Calculate the length of each side of the triangle.

25 Bob is twice as old as Megan. In six years, their ages will total 60 years.
How old is each of them now?

26 Tom is four years older than Mike. Nine years ago, Tom's age was five times Mike's age.
How old is each of them now?

27 The sum of two numbers is 84. The first number is nine more than four times the second number.
Calculate the two numbers.

28 Find three consecutive integers that have a sum of 48.

29 Find two consecutive even integers that have a sum of (−66).

30 Find three consecutive odd integers so that the sum of the smallest number and four times the biggest number is equal to 61.

Unit 3 Ratio, proportion and percentage

Key vocabulary

direct proportion	per cent	percentage increase
equivalent ratio	percentage	ratio
inverse proportion	percentage decrease	

A Ratio

Often, we need to **compare** different numbers or quantities.

For example, we might want to compare how many girls and how many boys there are in a class. We could just say 'There are 22 girls and 10 boys in this class', but in maths we like to use symbols to make things shorter and easier to write.

The symbol we use when we compare things like this is ':'. It is read as 'to' or 'is to'. So we could say that the number of girls to boys in the class is 22:10 (say '22 to 10'). This is called a ratio.

A ratio is always written in a special **order** – and we must tell everyone what this order is. So the ratio of **girls to boys** is 22:10, but the ratio of **boys to girls** is 10:22.

A ratio is written to compare things in one particular 'story'. We have compared the number of boys and girls in a particular class. The ratio of boys to girls in another class (or school, or town, etc.) will probably be different, and so we must tell everyone the story connected to any ratio we write.

We use a ratio when we compare different numbers or amounts of the same kind of things.

● We have compared the numbers of **girls** and **boys** – girls and boys are different but they are both kinds of **children** (or **pupils**)!
● We can compare the ages of two people, but we cannot compare the age of one person and the height of another.
● The length and width of a rectangle can be compared in a ratio because they are both measurements. If we compare different measurements we must make sure the measurements are in the **same units** before we can write it as a ratio.

A ratio does not include units (they must be the same, so we can leave them out).

Example

A father is 1.78 m tall and his son is 143 cm tall.
Write their heights as a ratio.

We cannot write a ratio with **metres** and **centimetres** mixed up. So, we must convert both measurements to the same units.
1.78 m = 178 cm
So the ratio of the heights of father to son is 178 : 143.

We can write a ratio to compare more than two things at the same time.

The father in the example above also has a daughter who is 125 cm tall. We can write the ratio of the heights of father to son to daughter as 178 : 143 : 125.

> A **ratio** is a useful way to write the **comparison** of two or more quantities that are the **same kind** of things or are in the **same units**.
> If we compare the two quantities a and b (in this order) we write this as $a : b$.

Examples

a) In a bus there are 35 men, 14 women and 21 children.
   Write these ratios.
   i) Men to women

   35 : 14

   ii) Women to men to children

   14 : 35 : 21

   iii) Men to total number of passengers

   The total number of passengers on the bus is 35 + 14 + 21 = 70.
   So the ratio is 35 : 70.

b) The bus fare in London for an adult is £1.
   The bus fare for a child is 75p.
   Write a ratio to compare the bus fares for child to adult.

   We cannot compare pounds (£) and pence (p), so we must change to the same units.
   £1 = 100p
   So the ratio of bus fares for child to adult is 75 : 100.

Ratios in their simplest form

Each number in a ratio is called a **term**. If the terms of a ratio have a **common factor**, we must divide **all the terms** by this factor. Dividing by the **Highest Common Factor (HCF)** will give us a ratio with the smallest possible **whole numbers** as terms. In other words, the ratio is in its **simplest form**.

When we write fractions in their simplest form, we divide the numerator and the denominator by their HCF. Ratios work similary.

Examples

Write each of the ratios from the examples above (on page 49) in its simplest form.

a) i) The terms of the ratio $35:14$ have an HCF of 7.
      So we must divide each term by 7.
      In its simplest form, the ratio is $5:2$.
   ii) The terms of the ratio $14:35:21$ also have an HCF of 7.
      In its simplest form, this ratio is $2:5:3$.
   iii) The terms of the ratio $35:70$ have an HCF of 35.
      In its simplest form, this ratio is $1:2$.
b) The terms of the ratio $75:100$ have an HCF of 25.
   In its simplest form, this ratio is $3:4$.

If a ratio is in its simplest form, the terms must be whole numbers.

Examples

Write each of these ratios in its simplest form.

a) $0.25:1.5$

   Multiply both terms by 100 so that they are whole numbers.
   $25:150$
   Now divide by the HCF, 25.
   $1:6$

b) $1.44:0.48$

   Multiply both terms by 100 so that they are whole numbers.
   $144:48$
   Now divide by the HCF, 48.
   $3:1$

c) $50.4:63$

   Multiply both terms by 10 so that they are whole numbers.
   $504:630$
   Now divide by the HCF, 126.
   $4:5$

Equivalent ratios

A ratio is used only to **compare** quantities. It does not give information about the actual values.

For example, if we know that a necklace is made using red beads and white beads in the ratio $3:4$, we know that for every three red beads in the necklace, there are four white beads – but it tells us nothing about the actual numbers of beads in the necklace.

Some of the **possible** numbers of beads are shown in the table on the right.

Red beads	White beads	Total beads
3	4	7
6	8	14
9	12	21
12	16	28
18	24	42

Can you see a pattern in the number of beads?
The number of red beads is a **multiple** of 3, the number of white beads is a **multiple** of 4 and the total number of beads is a **multiple** of 7.

So, we can use any of the ratios $3:4$ or $6:8$ or $9:12$ or $12:16$ or $18:24$, etc. to describe the ratio of red beads to white beads (but only the ratio $3:4$ is in its simplest form).

All these ratios are called equivalent ratios. These are very similar to **equivalent fractions**.

Equivalent ratios are **not** in their simplest form.

In their simplest form, equivalent ratios can all be written as the same ratio.

Exercise 1

1 Write each of these ratios in its simplest form.

a) $65:30$ b) $144:128$ c) $36:132$
d) $136:51$ e) $192:75$ f) $135:240$
g) $418:242$ h) $308:220$ i) $560:175$
j) $162:384$

2 Write each of these ratios in its simplest form.

a) 0.09 : 0.21   b) 0.84 : 1.12   c) 0.192 : 0.064
d) 1.8 : 0.4   e) 0.63 : 9.45   f) 1.26 : 0.315
g) 7.56 : 14.04   h) 19.747 : 4.9   i) 0.84 : 6.846
j) 6.4 : 9.6 : 16

3 Write the ratio of the first quantity to the second in its simplest form.

a) 45p to £1   b) 25 cm to 1.25 m
c) 0.25 km to 75 m   d) 0.2 kg to 40 g
e) 375 m to 1 km   f) 35 minutes to 1 hour
g) 15 mm to 2 cm   h) 1250 m^2 to 1 ha
i) 48 ha to 1 km^2 j) 0.6 minutes to 27 seconds
k) 3.2 hours to 72 minutes   l) £3 to 75p
m) 6 hours to 2 days   n) 9 months to 2 years
o) 0.035 litres to 105 ml   p) 5.04 kg to 360 g
q) 4.2 m to 6 m   r) 165 g to 3 kg
s) 2.25 hours to 90 minutes   t) 1.11 litres to 2590 ml

4 A shop sells 60 pastries, 45 pies and 80 sausage rolls.
Write down each of these ratios in its simplest form.

a) The number of sausage rolls to the number of pies
b) The number of pastries to the number of sausage rolls
c) The number of pastries to the number of pies

5 For her son's birthday party, Mrs Robinson buys 15 hats, 36 balloons and 12 toys.
Write down each of these ratios in its simplest form.

a) The number of balloons to toys
b) The number of toys to hats
c) The number of hats to balloons

6 In a Year 8 class there are 36 pupils. 20 of the pupils are under 13 years old.
Write down each of these ratios in its simplest form.

a) The number of pupils under 13 to the total number of pupils
b) The number of pupils over 13 to the total number of pupils
c) The number of pupils under 13 to the number of pupils over 13 years old

7 80 cars start a race. 64 of these cars finish the race.
Write down each of these ratios in its simplest form.

a) The number of cars that finished to the total number of cars
b) The number of cars that did not finish to the total number of cars
c) The number of cars that finished to the number of cars that did not finish

8 Copy these ratios. Write a number in each box to make the ratios equivalent.

a) i) $2:3 = 4:\square$ ii) $2:3 = \square:9$ b) i) $5:1 = 10:\square$ ii) $5:1 = \square:15$

iii) $2:3 = 10:\square$ iv) $2:3 = \square:21$ iii) $5:1 = 40:\square$ iv) $5:1 = \square:10$

c) $3:9 = 4:\square$ d) $2:\square = 3:15$

e) $4:3 = \square:6$ f) $5:11 = 10:\square$

g) $\square:7 = 15:4$ h) $\square:5.7 = 8:12$

i) $14:9 = 7:\square$ j) $0.84:\square = 0.6:0.5$

k) $3.2:0.2 = \square:1.1$ l) $12:25 = \square:5$

9 Change each of these ratios to equivalent ratios with whole numbers.

a) $0.8:0.6$ b) $2.5:2$ c) $0.2:1$

d) $3.2:1.2$ e) $\frac{1}{4}:3$ f) $\frac{2}{5}:\frac{3}{5}$

g) $\frac{3}{10}:\frac{7}{20}$ h) $2\frac{1}{3}:5$ i) $0.3:2:0.7$

j) $1\frac{1}{2}:2\frac{1}{2}:3$

10 The heights of two friends are in the ratio $7:9$. The shorter of the friends is 154 cm tall.
What is the height of the taller friend?

11 Sugar and flour are mixed in the ratio $2:3$.
How much sugar is mixed with 600 g of flour?

12 The ratio of boys to girls in a school is $4:5$. There are 80 girls.
How many boys are there?

13 Birget earns \$800 per month. Erik earns \$720 per month.
Their salaries are in the same ratio as their ages. Birget is 20 years old.
How old is Erik?

14 A necklace contains 30 black beads and 45 gold beads.
A second necklace has the same ratio of black beads to gold beads.
It has 44 black beads.
How many gold beads does it have?

Using ratios – dividing or sharing quantities in a given ratio

Sometimes we need to share a certain number of things between groups using a given ratio.

a) Philip and Sue are given £143 by their grandmother. Philip is six years old and Sue is five years old. They must share the money in the same ratio as their ages.
How much money does each of them get?

The ratio of their ages is $6:5$.
This means that Philip gets **six parts** and Sue gets **five parts** of the money.
So we must divide the money into $6 + 5 = 11$ equal parts.
Philip gets 6 out of 11 parts $= \frac{6}{11}$ and Sue gets 5 out of 11 parts $= \frac{5}{11}$.
Philip gets $\frac{6}{11} \times £143 = 6 \times £13 = £78$ and
Sue gets $\frac{5}{11} \times £143 = 5 \times £13 = £65$

b) A box contains red, white and blue buttons in the ratio $1:2:5$.
There are 104 buttons in the box. How many of them are blue?

We must divide the buttons in the ratio of $1:2:5$.
We have a total of $1 + 2 + 5 = 8$ equal parts.
5 of the 8 parts are blue buttons, so $\frac{5}{8} \times 104 = 65$ buttons are blue.

c) An alloy is made of the metals tin and zinc. $\frac{4}{10}$ of the alloy is tin.
What is the ratio of tin to zinc in its simplest form?

We know that 4 parts out of a total of 10 parts in the alloy is tin $\left(\frac{4}{10}\right)$.
This means that the other 6 parts must be zinc.
So the ratio of tin to zinc is $4:6 = 2:3$ in its simplest form.

d) A piece of string is 30 cm long. It is divided into two smaller pieces in the ratio of $1:5$.
Find the length of each of the two smaller pieces of string.

We must divide the string in the ratio of $1:5$.
We have a total of $1 + 5 = 6$ equal parts.
One piece is $\frac{1}{6}$ of the total length. $\frac{1}{6} \times 30\,\text{cm} = 5\,\text{cm}$.
The other piece is $\frac{5}{6}$ of the total length. $\frac{5}{6} \times 30\,\text{cm} = 25\,\text{cm}$.

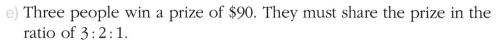

e) Three people win a prize of $90. They must share the prize in the ratio of 3 : 2 : 1.
Calculate
  i) the smallest share
  ii) the biggest share.

We have a total of 3 + 2 + 1 = 6 equal parts.
  i) The smallest share is $\frac{1}{6}$ × $90 = $15.
  ii) The biggest share is $\frac{3}{6}$ × $90 = $45.

f) Aromatico Brand coffee powder is a mixture of Grade A and Grade B coffee powder in the ratio of 2 : 3. We know that 1 kg of Grade A costs $12 and 1 kg of Grade B costs $9.
What is the cost of 1 kg of the Aromatico Brand mixture?

If the ratio of the mixture is 2 : 3, we have a total of 2 + 3 = 5 equal parts.
So Grade A is $\frac{2}{5}$ of the mixture. This means that $\frac{2}{5}$ of 1 kg is Grade A.
$\frac{2}{5}$ of 1 kg of Grade A costs $\frac{2}{5}$ × $12 = $4.80.
Grade B is $\frac{3}{5}$ of the mixture. This means that $\frac{3}{5}$ of 1 kg is Grade B.
$\frac{3}{5}$ of 1 kg of Grade B costs $\frac{3}{5}$ × $9 = $5.40.

The total cost of 1 kg of Aromatico Brand coffee is
$4.80 + $5.40 = $10.20

 Exercise 2

1 Share each of the following quantities in these ratios.
  i) 3 : 1
  ii) 1 : 7
  iii) 4 : 2

  a) 48 marbles
  b) 24 sweets
  c) 80 kg
  d) 288 ml
  e) £1200

2 A box contains blue pens and red pens. $\frac{3}{11}$ of the pens are red.
  What is the ratio of blue pens to red pens in its simplest form?

3 A pond contains 40 fish. $\frac{2}{5}$ of the fish are salmon. The rest of the fish are trout.
   Find the ratio of salmon to trout in its simplest form.

4 A box contains gold coins and silver coins. The ratio of gold coins to silver
   coins is $1:9$. There are 20 coins in the box.
   How many silver coins are in the box?

5 David and Sarah share £48 in the ratio $3:1$.
   How much money does each of them get?

6 In a school, the ratio of the number of boys to the number of girls is $3:5$.

   a) What fraction of the pupils are girls?
   b) If there are 376 pupils in the school, how many boys are there?

7 Jean-Paul is 13 years old and Michelle is 12 years old. They share €1025 in
   the ratio of their ages.

   a) What fraction of the money does Jean-Paul get?
   b) How much money does Michelle get?

8 $480 is shared in the ratio $7:3$.
   What is the difference between the bigger share and the smaller share?

9 In 1905, the total population of England and Scotland was 32 528 000.
   The ratio of the population of England to the population of Scotland was $15:1$.
   What was the population of Scotland in 1905?

10 In the UK there are 240 939 km² of land. The ratio of agricultural land to
   non-agricultural land is about $7:3$.
   Estimate the area of land used for agriculture.

11 At the start of a game, Gretchen and Viktor have 40 cards each.
   At the end of the game, the number of cards that Gretchen and Viktor have
   is in the ratio $3:5$.

   a) How many cards do Gretchen and Viktor have at the end of the game?
   b) How many cards did Viktor win from Gretchen in the game?

12 The lengths of the sides of a triangle are in the ratio $4:6:9$.
   The total length of the sides is 38 cm.
   Calculate the length of each side.

13 To make concrete, a builder mixes stone, sand and cement in the ratio $4:2:1$.
   The builder wants 350 kg of concrete.
   How much stone does he need?

14 The angles in a triangle are in the ratio $2:3:4$.
   Calculate each of the angles.

15 Yorkshire Tea is made by mixing 1st Grade and 2nd Grade tea leaves in the ratio of $1:4$.
1 kg of 1st Grade tea costs £8.
1 kg of 2nd Grade tea costs £6.
How much does 1 kg of Yorkshire Tea cost?

16 Ruby Brand Oil is made from a mixture of Oil A and Oil B in the ratio of $1:3$.
100 ml of Oil A costs $4.
100 ml of Oil B costs $2.
How much does 1 litre of Ruby Brand Oil cost?

17 A sum of money is divided in the ratio $3:5:9$.
Calculate the smallest share if the biggest share is:

a) €369
b) €558.

18 In a quadrilateral, *ABCD*, the angles *A*, *B* and *C* are in the ratio $1:3:7$.
Calculate $\hat{C}$ if the sum of $\hat{A}$, $\hat{B}$ and $\hat{C}$ is:

a) 242°
b) 253°.

19 A sum of money is divided among three people in the ratio $15:18:7$.
Find

i) the total sum of money
ii) the biggest share

if the smallest share is:

a) £133
b) £217.

20 $160 is shared among three people.
Calculate the difference between the smallest share and the biggest share for each of these ratios:

a) $2:5:13$
b) $19:18:3$

21 Some money is divided among Peter, Paul and Jane in the ratio $13:12:7$.
Calculate how much Paul gets if the difference between Peter's money and Jane's is:

a) £90
b) £360.

B Direct proportion

Sometimes we need to compare quantities of things that are **different**. We can use a **ratio** only to compare quantities that are of the **same kind** of things.

This table shows the costs of buying different numbers of footballs.

Number of footballs	Cost of footballs
1	$2
2	$4
3	$6
4	$8
5	$10

increasing number (left)   increasing cost (right)

You can see that the cost **increases** by $2 for every extra football we buy.
If the number of footballs is doubled (×2), then the cost will also be doubled.

This means that the number of footballs and the cost are **linked** and **change together at the same rate**.

Let us look at some ratios. Remember that we can use a ratio only to compare the same sort of things. So we cannot use a ratio to compare the number of footballs and the cost, but we can compare footballs with footballs, and cost with cost.

If we compare four footballs to two footballs, we get a ratio of $4:2 = 2:1$.
If we compare the cost of four footballs to the cost of two footballs, we get a ratio of $8 to $4 or $8:4 = 2:1$.

In the same way, if we compare three footballs to one football, we get a ratio of $3:1$.

If we compare the cost of three footballs to the cost of one football, we get a ratio of $6 to $2 or $6:2 = 3:1$

And so on …

From this we can see that the footballs and cost (which are two different things) **change in the same ratio**. We say that they change in direct proportion.

> Quantities that increase or decrease in the same ratio are in direct proportion.

To solve problems involving direct proportion, it often helps to ask yourself these four questions:

- What must we work out?
- What do we know?
- What is the cost (or whatever we need to find) of one of the things?
- What is the cost (or whatever we need to find) we need to know?

Examples
a) Four cakes cost 120 baht. How much do seven cakes cost?

- What must we work out?	The cost of seven cakes.
- What do we know?	Four cakes cost 120 baht.
- What is the cost of one cake?	So one cake costs $120 \div 4 = 30$ baht.
- What is the cost of seven cakes?	So seven cakes cost $30 \times 7 = 210$ baht.

b) 3 m of wooden fencing costs $24. How much will Alan pay for 20 m?

- What must we work out?	The cost of 20 m of fencing.
- What do we know?	3 m of fencing costs $24.
- What is the cost of 1 m?	So 1 m of fencing costs $24 \div 3 = \$8$.
- What is the cost of 20 m?	So 20 m of fencing costs $\$8 \times 20 = \$160$.

Of course, you don't have to write out the four questions every time.

Example
Find the cost of a 13 kg box of biscuits if a 6 kg box costs $27.

6 kg of biscuits cost $27.
So 1 kg of biscuits costs $27 \div 6 = \$4.50$.
So 13 kg of biscuits cost $\$4.50 \times 13 = \$58.50$.

In the examples above, we write down the **information we know** so that the type of thing we must work out (e.g. cost or quantity) is at the **end** of the statement. This will help us with the calculation. In the last example above, we must work out the **cost** of 13 kg of biscuits. So we write down what we know as:

6 kg of biscuits **cost** $27.

Sometimes the information given in the question is not in the same order as we need for the calculation. We must then turn the information we are given around and write it so that the type of thing we must work out is at the end of the statement.

Example
: A car uses 21 litres of petrol to travel 315 km. How much petrol will it need to travel 90 km?

- What must we work out?   How many **litres of petrol** to travel 90 km?
- What do we know?   To travel 315 km a car uses 21 **litres of petrol**.

So to travel 1 km a car uses $21 \div 315 = \frac{1}{15}$ litres.

So to travel 90 km a car uses $90 \times \frac{1}{15} = 6$ litres.

Exercise 3

1 Five candles cost 80 pence.
What is the cost of eight candles?

2 Ben works for four hours and earns £24.
How much does he earn if he works for 10 hours?

3 Amelie pays €1.90 for two cups of coffee.
How much will she pay for three cups of coffee?

4 A car travels 6 km in 8 minutes.
How far will the car travel in 20 minutes if it travels at the same speed?

5 Four English dictionaries cost £13.20. How much will 12 dictionaries cost?

6 12 kg of bananas cost $16.00.

   a) Calculate the cost of 18 kg of bananas.
   b) How many kilograms of bananas can I buy for $20?

7 This recipe will make 20 scones.

   a) How much dried fruit is needed to make 50 scones?
   b) How much butter is needed to make 12 scones?

| 500 g flour |
| 250 g butter |
| 100 g dried fruit |
| water |

8 This recipe makes apple crumble for six people.

   a) How much sugar is needed to make apple crumble for four people?
   b) How much apple is needed to make apple crumble for eight people?

| 540 g apples |
| 75 g butter |
| 150 g flour |
| 75 g sugar |

9 1400 Thai baht is about the same as £20. Sue spends £47.
How many Thai baht is this?

10 Mary phones her uncle in New York. A five-minute call to New York
costs £1.10.

a) How much does a seven-minute call to New York cost?
b) Mary's call cost £2.64. How many minutes did she speak for?

11 A car travels 6 miles in 9 minutes. If the car travels at the same speed,

a) how long will it take to travel 8 miles
b) how far will it travel in 24 minutes?

12 5 litres of paint can cover an area of $30 \, m^2$.

a) What area will 2 litres of paint cover?
b) How much paint will cover $72 \, m^2$?

13 A school is taking three different
groups of people to the zoo.

a) What is the total cost of
Tuesday's trip?
b) How many people are going on
Wednesday?

Monday:	45 people are going.
	The total cost is $234.
Tuesday:	25 people are going.
Wednesday:	The total cost is $166.40

14 Mr Kahn is filling a small pool for his children to play in. The pool holds
42 litres of water.
Every 5 minutes, 12 litres of water is added to the pool until it is full.
How long will it take to fill the pool?

15 A man plans to travel 1273 km. Petrol costs $1.18 per litre.
His car travels 19 km on 1 litre of petrol.
How much will he spend on petrol for the trip?

16 It costs nine tourists €1620 to stay at a hotel for four days.
a) How much would it cost 15 tourists to stay for six days?
b) How many days can 10 tourists stay at the hotel if they have €2250?

17 A sheet of brass is a rectangle with a length of 1.5 m and a width of 75 cm.
It weighs 7.2 kg.
Find the area of another similar brass sheet that weighs 12.8 kg.

C Inverse proportion

Sometimes we find that quantities change in the same ratio, but as one of them gets **bigger**, the other one gets **smaller**. The quantities are still **in proportion**, but this time the relationship is 'upside-down'.

This table shows the speed of a car and the time it takes to travel 120 km.

Speed of car in km/h	Time in hours
20	6
30	4
40	3
60	2
120	1

increasing speed

decreasing time

We can see from the table that, as the car goes faster (speed goes up), it takes less time to travel the same distance (the time taken goes down).

The English word for 'upside-down' is **invert** or **inverse**. So, we call this 'upside-down' proportion inverse proportion.

> If one quantity increases at the same rate that another decreases, they are inversely proportional to each other.

We can use the same four questions as before to help us solve problems involving inverse proportion. You don't need to write the questions down, but you might find it helpful to do so at first.

Examples

a) Ten men can finish some work in six days.
   How many days will it take 15 men to do the same amount of work?

- What must we work out?   How many days it will take 15 men.
- What do we know?   Ten men can do the work in six days.

Now we must think: will one man do all the work in more days or fewer days? More, of course!

- How long would 1 man take?   One man could do the work in
  $6 \times 10 = 60$ days.

Again, we must think: will 15 men do all the work in more days or fewer days? Fewer!

- How long would 15 men take?   15 men could do the work in
  $60 \div 15 = 4$ days.

b) It takes a car 20 minutes to travel from Town A to Town B at 60 km/h.
At what speed must the car travel to do the same journey in 15 minutes?

- What must we work out?   What speed the car must travel for 15 minutes.

- What do we know?   The journey takes 20 minutes at a speed of 60 km/h.

Now we must think: to do the journey in 1 minute, must the car travel faster or slower? Faster!

- What is the speed for 1 minute?   The journey would take 1 minute at a speed of $60 \times 20 = 1200$ km/h.

Again, we must think: to do the journey in 15 minutes, must the car travel faster or slower? Slower!

- What is the speed for 15 minutes?   The journey would take 15 minutes at a speed of $1200 \div 15 = 80$ km/h.

 Exercise 4

1 Six workers can paint a factory in four days.
How many workers will be needed to finish the job in three days?

2 A farmer has 90 sheep. He has enough food to feed the sheep for 60 days.
How many sheep must he sell so that the food will last for 100 days?

3 Nine workers can dig a hole in ten days.
How many days will it take for six workers to dig the same hole?

4 A car travels at 70 km/h to complete a journey in 54 minutes.
If it travels at 45 km/h, how long will the same journey take?

5 If a ball moves at 135 metres per second it will hit a wall in 4 seconds.
If it moves at 120 metres per second, how long will it take to hit the same wall?

6 It takes four hosepipes 70 minutes to fill a swimming pool.
How long will it take if seven hosepipes are used?

7 A school has enough money to buy eight books that cost $5.50 each.
If they buy books that cost $8.80, how many can they buy for the same amount of money?

8 Ten cooks each work for eight hours to prepare a meal for 536 people.
How many cooks will you need to prepare the same meal for 737 people if they can only work for five hours each?

9 A builder employs 50 men to work eight hours each day to build a road that is 3000 m long.

He promises to finish the road in 30 days. After 20 days, only 1200 m is finished.

How many men must work to finish the road on time if each man works 10 hours per day?

D Percentage

We can see advertisements like these every day in newspapers, magazines, on the internet and on TV.

The word per cent comes from the Latin 'per centum' which means 'out of one hundred'.

The symbol we use for per cent is **%**. This symbol used to be written as /100 but when people had written it many times, it slowly changed into the symbol that we use now!

So, 7% is read as '**seven per cent**' and it means 'seven parts out of one hundred'.

Changing a percentage into a fraction or a decimal

We know that 7% means 'seven parts out of one hundred'.

So we can write it as a fraction, $\frac{7}{100}$, and as a decimal, 0.07.

This means that 7% $= \frac{7}{100} = 0.07$.

> Every percentage can be written as a **fraction** with **100 as the denominator**.
> This means that every **percentage** can also be written as a **decimal**.

Examples

a) Write each of these percentages as a fraction in its simplest form.

i) $19\% = \dfrac{19}{100}$ (This is already in its simplest form.)

ii) $40\% = \dfrac{40}{100} = \dfrac{4}{10} = \dfrac{2}{5}$

iii) $62.5 = \dfrac{62.5}{100} = \dfrac{625}{1000} = \dfrac{5}{8}$

iv) $17\frac{1}{2}\% = \dfrac{17\frac{1}{2}}{100} = \dfrac{\frac{35}{2}}{100} = \dfrac{35}{2} \times \dfrac{1}{100} = \dfrac{35}{200} = \dfrac{7}{40}$

v) $130\% = \dfrac{130}{100} = \dfrac{13}{10} = 1\frac{3}{10}$

vi) $350\% = \dfrac{350}{100} = \dfrac{35}{10} = \dfrac{7}{2} = 3\frac{1}{2}$

b) Write each of these percentages as a decimal.

i) $20\% = \dfrac{20}{100} = 0.2$

ii) $7\frac{1}{2}\% = 7.5\% = \dfrac{7.5}{100} = \dfrac{75}{1000} = 0.075$

iii) $140\% = \dfrac{140}{100} = 1.4$

iv) $13\frac{3}{5}\% = 13.6\% = \dfrac{13.6}{100} = \dfrac{136}{1000} = 0.136$

Changing a decimal into a percentage

To change a decimal into a percentage, it is easy to write the decimal as a **fraction with a denominator of 100**. By definition, the numerator of this fraction will be the percentage.

Examples

Write each of these decimals as a percentage.

a) $0.27 = \dfrac{27}{100} = 27\%$

b) $0.82 = \dfrac{82}{100} = 82\%$

c) $0.03 = \dfrac{3}{100} = 3\%$

d) $0.746 = \dfrac{74.6}{100} = 74.6\%$

e) $1.519 = \dfrac{151.9}{100} = 151.9\%$

f) $4.0 = \dfrac{400}{100} = 400\%$

Changing a fraction into a percentage

We know that any percentage can be written as a fraction with a denominator of 100.

So, to write any fraction as a percentage, we must first write an **equivalent fraction** with a denominator of 100 to find the percentage (numerator).

If we have a fraction $\frac{a}{b}$ then it can be written as some percentage p.

Therefore $\frac{a}{b} = \frac{p}{100}$. If we multiply both sides by 100 we find the percentage:

$$p = \frac{a}{b} \times 100\%$$

This is also true for all decimals, so we have the rule:

> To change any fraction or decimal into a percentage, multiply it by 100.

Examples

Write each of these fractions as a percentage.

a) $\frac{3}{5} = \frac{3}{5} \times 100\% = 60\%$

b) $\frac{7}{25} = \frac{7}{25} \times 100\% = 28\%$

c) $\frac{9}{40} = \frac{9}{40} \times 100\% = 22\frac{1}{2}\%$

d) $2\frac{2}{3} = \frac{8}{3} \times 100\% = \frac{800}{3}\% = 266\frac{2}{3}\%$

Exercise 5

1 Write each of these percentages as a fraction in its simplest form.

a) 210%   b) 140%   c) 4.8%
d) 88%   e) 18%   f) 2.5%
g) 35%   h) 0.25%   i) $1\frac{1}{3}\%$
j) $5\frac{2}{5}\%$

2 Write each of these percentages as a decimal.

a) 7%   b) 46%   c) 99%
d) 300%   e) 480%   f) 0.68%
g) 0.04%   h) 1.002%   i) $2\frac{4}{5}\%$
j) $14\frac{1}{4}\%$

3 Write each of these fractions as a percentage.

a) $\frac{1}{2}$ b) $1\frac{1}{4}$ c) $1\frac{3}{10}$

d) $2\frac{2}{5}$ e) $3\frac{3}{8}$ f) $\frac{11}{20}$

g) $\frac{12}{15}$ h) $\frac{13}{25}$ i) $\frac{19}{50}$

j) $\frac{25}{4}$ k) $\frac{5}{16}$ l) $\frac{9}{125}$

m) $\frac{3}{64}$ n) $\frac{4}{625}$ o) $1\frac{1}{3}$

4 Write each of these decimals as a percentage.

a) 0.07   b) 0.058   c) 0.14

d) 0.39   e) 0.136   f) 0.505

g) 0.027   h) 0.5218   i) 2.43

j) 6.325   k) 100   l) 0.0011

m) 8.9   n) 4   o) 1.25

5 Copy and complete this table.

Percentage	Fraction	Decimal
	$\frac{3}{5}$	
11%		
		0.175
		0.095
$78\frac{1}{2}\%$		

Comparing fractions

We have already learned a way to compare fractions with different denominators:

- First calculate the **lowest common multiple (LCM)** for all the denominators.
- Write each fraction as an **equivalent fraction** with the **same denominator** (LCM).
- Compare the **numerators** of the equivalent fractions.

A much quicker way to compare fractions is to change them all into percentages first. Then it is easy to see which is the biggest, which is the smallest and so on.

Example

Which fraction is bigger, $\frac{17}{20}$ or $\frac{21}{25}$?

Write both fractions as percentages.

$$\frac{17}{\cancel{20}} \times \cancel{100}^{5}\% = 17 \times 5\%$$

$$= 85\%$$

$$\frac{21}{\cancel{25}} \times \cancel{100}^{4}\% = 21 \times 4\%$$

$$= 84\%$$

So $\frac{17}{20}$ is a bigger fraction than $\frac{21}{25}$.

Writing one number as a percentage of another

To work out one number as a percentage of another, there are three easy steps:

1 Make sure both quantities are in the **same units**.
2 Write one quantity as a **fraction** of the other.
3 Multiply this fraction by 100 to change it into a percentage.

Examples

a) What percentage is 24 out of 40?

Fraction: $\frac{24}{40}$

Percentage: $\frac{24}{40} \times 100\% = 60\%$

b) There are 40 pupils in a class. 12 of the pupils are in a football team. What percentage of the pupils are not in the football team?

Number of pupils not in the football team is $40 - 12 = 28$

Fraction: $\frac{28}{40}$

Percentage: $\frac{28}{40} \times 100\% = 70\%$

c) Write 12 minutes as a percentage of 1 hour.

Change one of the quantities so that both are in the same units:
1 hour = 60 minutes

Fraction: $\frac{12}{60}$

Percentage: $\frac{12}{60} \times 100\% = 20\%$

If we write a larger number as a percentage of a smaller number, the answer will be more than 100%.

Example

Write 32.5 m as a percentage of 250 cm.

Change one of the quantities so that both are in the same units:
32.5 m = 325 cm

Fraction: $\frac{325}{250}$

Percentage: $\frac{325}{250} \times 100\% = 130\%$

 Exercise 6

1 Write the first value as a percentage of the second.

a) £14, £42
b) 33 cm, 3.96 m
c) 225 mm, 20 cm
d) $4.40, 99¢
e) 45 kg, 36 kg
f) 75 cm, 6 m
g) 6 hours, 1 day
h) 20 minutes, 1 hour
i) 175 ml, 1 litre
j) 1 km, 300 m
k) 48p, £1.44
l) 2 years, 18 months

2 Rachel played 12 games of squash. She won nine of these games.
What percentage of the games did she win?

3 In a spelling test, Hans correctly spelled 19 out of 20 words.
What percentage did he spell correctly?

4 This table shows the scores that David gained in his final examinations.
Calculate the percentage scores that David gained for each of the subjects.

	Subject	Marks
a)	English	90 out of 100
b)	Mathematics	64 out of 80
c)	Science	60 out of 75
d)	Computer Studies	78 out of 120
e)	Social Studies	77 out of 140

5 In a school cross-country race, 72 out of 80 pupils finished the race.

a) What percentage of pupils finished?
b) What percentage of pupils did not?

6 A gardener found that 63 out of 90 lettuces he planted grew well.
What percentage of lettuces did not grow well?

7 Kofi and Manu both wanted to be captain of the football club.
There are 25 members in the club. 18 members voted for Kofi.

a) What percentage voted for Kofi?
b) What percentage voted for Manu?

8 Ekua does two jobs and earns $28 per day. She earns $21 for delivering newspapers. She earns $7 for mowing lawns.
a) What percentage of her money comes from delivering papers?
b) What percentage of her money comes from mowing lawns?

9 Write these numbers in ascending order.
$\frac{17}{40}$ 0.42 $\frac{9}{20}$ 43%

10 Write these numbers in descending order.
$\frac{23}{80}$ 28% $\frac{57}{200}$ 0.2805

11 In a badminton competition, team A won eight out of the eleven games they played. Team B won five of their seven games.
Which team played better overall?

12 Norman earns £4.50 per hour. His wages rise by 27 pence per hour.
What is his percentage wage rise?

13 A new car costs €13 500. The dealer gives a discount of €1282.50.
What is the percentage discount?

14 There are 600 pupils in Years 7, 8, 9, 10 and 11 of a school. 360 pupils are in Years 8 and 9. 15% of the pupils are in Years 10 and 11.
What percentage of the pupils are in Year 7?

Finding a given percentage of a quantity

To work out a given percentage of any quantity is quite easy.

1 Write the percentage as a fraction (remember, % means 'out of 100').
2 Multiply the quantity by this fraction.

NOTE: Remember that 'of' means multiply in maths.

Examples

a) Find 12% of $250.

$12\% = \frac{12}{100}$

So $\$\left(\frac{12}{100} \times 250\right) = \$30$

b) There are 40 pupils in a class. 45% of them are boys.
How many are girls?

$$\text{Number of boys} = 45\% \text{ of } 40$$
$$= \frac{45}{100} \times 40$$
$$= 18$$
$$\text{Number of girls} = 40 - 18$$
$$= 22$$

Exercise 7

1 Calculate these percentages.
  a) 10% of 90
  b) 6% of 200
  c) 38% of 400
  d) $10\frac{1}{2}$ % of 500
  e) 86% of 35
  f) 13.25% of 10 000
  g) 150% of 754
  h) 2% of 124
  i) 16% of 350
  j) 0.25% of 4 000
  k) 120% of 600
  l) 30% of 1 170

2 Calculate these percentages.
  a) $66\frac{2}{3}$% of 72 litres
  b) 45% of 4 kg
  c) 12% of 95 km
  d) 7.5% of £2500
  e) $37\frac{1}{2}$% of 56 cm
  f) $33\frac{1}{3}$% of 48 m
  g) 20.6% of 5000 people
  h) 135% of $300
  i) $112\frac{1}{2}$% of 400 g

3 There are 450 seats in a theatre. 60% of the seats are downstairs.
How many seats are there downstairs?

4 Alain invests $300 at a bank. He earns 6% interest per year.
How much interest does he get in one year?

5 Irina gets a 15% discount on a theatre ticket because she is a student.
The normal price of the ticket is $18.
How much money does she save?

6 In one town there are 8500 voters. On election day, 15% of them did not vote.
How many people did vote?

7 Soya beans contain about 39.5% protein. I have 120 kg of soya beans.
How many kilograms of protein do my soya beans contain?

8 There are 350 people living in a village.
30% of the people are men and 32% of them are women.
  a) What is the percentage of children in the village?
  b) How many women are in the village?

9 A rock weighs 150 g. A scientist finds that $37\frac{1}{2}$% of the rock is copper. How many grams of copper are in the rock?

10 Mrs Walsh needs to pay an 18% deposit on a new TV. The TV costs £600. How much deposit must she pay?

Percentage increase and decrease

Sometimes a value is increased or decreased by a given percentage. We can work out the **new amount** by following these two easy steps:

1 Calculate the value of the change.
2 a) If the change is an **increase**, then **add** this amount to the **old amount** to give the new one.
   b) If the change is a **decrease**, then **subtract** this amount from the **old amount** to give the new one.

Examples

a) A shirt normally costs $24. It is reduced by 15% in a sale. What is the new price of the shirt in the sale?

● Calculate the value of the reduction.   15% of $24 = $$\left(\frac{15}{100} \times 24\right)$
$$= \$3.60$$

● Subtract this from the old price. $24.00 − $3.60 = $20.40

The new price of the shirt in the sale is $20.40.

b) A packet of rice weighs 440 g. A special offer gives a packet with 30% more. How much does the special offer packet weigh?

● Calculate the value of the increase.   30% of 440 g = $\left(\frac{30}{100} \times 440\right)$ g = 132 g

● Add this to the old weight.   440 g + 132 g = 572 g

The weight of the special offer packet is 572 g.

We can also calculate the percentage increase/decrease if we are given both the **old** and **new** amounts:

1 Calculate the value of the change.
2 Write the change as a fraction of the **old amount**.
3 Change the fraction to a percentage.

Example

During the year 2001 the population of a city increased from 1 560 000 to 3 510 000.

Find the percentage increase in population.

- Calculate the value of the increase. $3\,510\,000 - 1\,560\,000 =$
    $1\,950\,000$

- Write the increase as a fraction $\frac{1\,950\,000}{1\,560\,000} = \frac{195}{156} = \frac{5}{4}$
    of the old amount.

- Change the fraction to a percentage. $\frac{5}{4} = \frac{5}{4} \times 100\% = 125\%$

The population increased by 125% during 2001.

Exercise 8

1   Calculate these percentage increases.

   a) $50 is increased by 60%
   b) £10 is increased by 30%
   c) €15 is increased by 10%
   d) $50 is increased by 15%
   e) 32 g is increased by $12\frac{1}{2}\%$
   f) 18 m² is increased by 5%
   g) 50 m is increased by 125%
   h) 69 litres is increased by 28%
   i) 48 km is increased by $33\frac{1}{3}\%$
   j) €2000 is increased by 0.5%
   k) 23 kg is increased by 300%
   l) 28 mm is increased by 75%

2   Calculate these percentage decreases.

   a) $600 is decreased by 15%
   b) €55 is decreased by 90%
   c) £42 is decreased by 20%
   d) €63 is decreased by 35%
   e) 64 g is decreased by $12\frac{1}{2}\%$
   f) 124 litres is decreased by 25%
   g) 225 m is decreased by 16%
   h) 428 kg is decreased by 35%
   i) 65 mm is decreased by 80%
   j) £450 is decreased by 1.5%
   k) 250 m² is decreased by 0.2%
   l) 175 km is decreased by 40%

3   Write the difference between the bigger number and the smaller number as a percentage of the first number.

   a) 100, 119
   b) 36, 45
   c) 90, 150
   d) 75, 100
   e) 100, 79
   f) 344, 301
   g) 96, 60
   h) 156, 351
   i) 450, 270
   j) 504, 1176
   k) 468, 351
   l) 336, 1134

4 A mobile phone company offers a 20% discount on calls made in March.
Calls normally cost 30 cents per minute.
How much do calls cost in March?

5 Robin earns £200 per week. He gets a pay rise of 7.5%.
What is his new weekly pay?

6 A packet of sugar contains 600 g. A special offer packet contains 15% extra.
How many grams of sugar are in the special offer packet?

7 A new car costs $12 000. After one year its value decreases by 20%.

a) What is its value at the end of one year?

During the second year its value decreases by 10%.

b) What is its value at the end of two years?

8 A new car costs $13 500. After one year its value decreases by 22%.

a) What is its value at the end of one year?

At the end of two years it was sold for $8200.

b) What percentage of its original value did the car lose in the second year?

9 A 5 litre can of paint covers 28 m². Jarrod buys three cans of paint to
cover 70 m².
What percentage of the paint does he use?

10 During 2008 the population of a village decreased from 323 to 260.
What is the percentage decrease in the population?

11 In 2007 a train can carry 360 people. In 2008 it can carry 8% more people.
In 2009 it can carry 8% more people than in 2008.
How many people can it carry in 2009?

12 John is 1.8 m tall. Ben is 8% taller than John. Wills is 10% shorter than John.
By what percentage is Ben taller than Wills?

13 The price of telephone calls is 20¢ per minute in 2008. In 2009 the cost is
increased by 15%, but businesses are given a 10% reduction on the new price.

a) How much do businesses pay for calls in 2009?
b) What is the percentage increase in the price of calls for businesses in 2009?

Working backwards with percentage increase/decrease problems

Sometimes we know the **percentage increase or decrease**, and we also know the **new amount**. We can then work out what the **old amount** was.

1 First we need to work out the new amount as a percentage. The **old amount** is 100%. So
   a) after an **increase**, the **new amount** is (100 + the increase)%
   b) after a **decrease**, the **new amount** is (100 − the decrease)%
2 Now we can use a **proportion calculation** to work out what the old (100%) amount was.

Examples

a) A shop sells videos with a 20% discount.
   Pam buys a video and pays £10.
   How much did this video cost before the discount?

   - Remember, the cost before the discount is 100%.
     Pam pays the cost after
     the discount.   Pam pays 100% − 20% = 80%
     She pays £10.   80% = £10
   - Now use a proportion   80% = £10
     calculation.   So 1% = $\frac{£10}{80}$

     So 100% = $£\left(\frac{10}{80} \times 100\right)$ = £12.50

   The cost of the video before the discount was £12.50.

b) Thierry gets a 5% pay rise. His new wage is $126 per week.
   What was Thierry's wage before his pay rise?

   - Remember, the wage before the rise is 100%.
     Thierry's new wage is after   Thierry's new wage is
     the pay rise.   100% + 5% = 105%
     He now gets $126.   105% = $126
   - Now use a proportion   105% = $126
     calculation.   So 1% = $\frac{\$126}{105}$

     So 100% = $\$\left(\frac{126}{105} \times 100\right)$

                           = $120

   Thierry's wage before the pay rise was $120.

Exercise 9

1 John saves 15% of his monthly salary. He saves £90 each month.
   What is his monthly salary?

2 Marie gets a 20% pay rise. Her new wage is €264 per week.
   What was Marie's wage before the rise?

3 At sports day, Nikolai jumps 1.8 m in the high jump.
   This is 5% lower than the best height Nikolai can jump.
   What is the best height he can jump?

4 In 2008 a house is worth $350 000.
   The value of this house increased by 12% during the year of 2007.
   What was the value of the house at the beginning of 2007?

5 30 grams of fruit provides 16.2 mg of vitamin C.
   This is 24% of the amount of vitamin C that a person needs in one day.
   How much vitamin C does a person need in one day?

6 A car is one year old. Its value decreased by 16%. It is now worth $13 440.
   How much was this car worth when it was new?

7 Alex gets a 3% increase in salary. His new salary is $1462.60 per month.
   What was Alex's salary before the rise?

8 When a shopkeeper opens a box of pears, he finds 10% of them are rotten
   and must be thrown away. He sells the 135 pears that are left.
   How many pears were in the box to start with?

9 The number of pupils in School A increases by 4% to 442.
   The number of pupils in School B decreases by 6% to 423.
   How many pupils were in School A and School B before the changes?

10 A man spends 88% of his salary. He now has £216 left.
   What is the man's salary?

11 An estate agency sold 20% more houses in 2008 than it did in 2007.
   In 2008, the agency sold 426 houses.
   How many houses did it sell in 2007?

12 A restaurant charges 12% for service. A customer pays a bill of €187.04
   What was the total of the bill before the service charge was added?

Unit 4

Applications of ratio, proportion and percentage

Key vocabulary

best buy	discount	loss
capital	down payment	marked price
compound interest	exchange rate	per annum
compounded	foreign currency	principal
cost price	hire purchase	profit
currency	instalment	sale price
currency converter	interest	selling price
deposit	interest rate	simple interest

A Ratio

Revision

We have learned that:

- A **ratio** is a useful way to **compare** two or more quantities that are the **same kind** of things.
- If we compare the two quantities a and b (in this order) we write this as $a : b$.
- All the quantities in a ratio must be in the **same units** before writing the ratio.
- When you write a ratio, there are **no units** of measurement.
- Ratios in their **simplest form** must always be written as **whole numbers**.

Examples

a) A cheese sandwich in London costs £2.25.
A can of soda costs 90p.
Write a ratio to compare the price of a can of soda to the price of a cheese sandwich.

$$90p : £2.25 = 90p : 225p$$
$$= 90 : 225$$
$$= 2 : 5$$

||||➡

b) Simplify the ratio $\frac{2}{3} : \frac{4}{9}$.

$$\frac{2}{3} : \frac{4}{9} = \frac{6}{9} : \frac{4}{9}$$

$$= \frac{6}{9} \times 9 : \frac{4}{9} \times 9$$

$$= 6 : 4 = 3 : 2$$

c) A box contains peanuts, cashew nuts and pecan nuts in the ratio $7 : 3 : 2$. There are 168 nuts in the box altogether.
How many cashew nuts are there?

Divide the nuts in the ratio of $7 : 3 : 2$.
Number of parts is $7 + 3 + 2 = 12$ equal parts

3 of the 12 parts are cashew nuts, so $\frac{3}{12} \times 168 = 42$ cashew nuts.

d) Three people win some prize money.
They must share the prize in the ratio of $3 : 2 : 1$.
The biggest share is $45.
Find the total prize money.

Divide the money in the ratio $3 : 2 : 1$.
Number of parts is $3 + 2 + 1 = 6$ equal parts
The biggest share is 3 of these 6 parts.

$$\frac{3}{6} \times \text{prize money} = \$45$$

$$\text{prize money} = \$45 \times \frac{6}{3} = \$90$$

B Proportion

Revision

Sometimes we need to compare quantities of things that are **different** and are measured in different units.

We have learned that:

- If both quantities increase (or decrease) at the same rate, they are in **direct proportion**.
- If one quantity increases while another quantity decreases at the same rate, they are **inversely proportional**.

Examples

a) 0.8 kg of fish costs $6.40.
  i) Calculate the cost of 2 kg of fish.

   0.8 kg of fish costs $6.40.
   1 kg of fish costs $6.40 ÷ 0.8 = £8.00.
   2 kg of fish costs $8.00 × 2 = $16.00.

ii) Calculate how many kilograms of fish I can buy for $9.60.

$6.40 will buy 0.8 kg of fish.
$1.00 will buy 0.8 ÷ 6.40 = 0.125 kg.
$9.60 will buy 0.125 × 9.60 = 1.2 kg.

b) Some emergency food will feed 30 men for 32 days.
   i) How long will this food feed 40 men?

30 men can eat for 32 days.
1 man can eat for 32 × 30 = 960 days.
40 men can eat for 960 ÷ 40 = 24 days.

   ii) If the food must last for 48 days, how many men will it feed now?

The food will last for 32 days if there are 30 men.
The food will last for 1 day if there are 30 × 32 = 960 men.
The food will last for 48 days if there are 960 ÷ 48 = 20 men.

C Percentage

Revision

We have learned that:

- The word **per cent** means 'out of one hundred'.
- 100% is '100 out of 100' or $\frac{100}{100} = 1$.
- To change a fraction or a decimal into a percentage, multiply it by 100.
- When a quantity changes, it is always 100% **before** the change.

Examples

a) Write each of these numbers as a percentage.

   i) $\frac{7}{20}$

$$\frac{7}{\overset{}{20}} \times \overset{5}{\cancel{100}}\% = 35\%$$

   ii) $2\frac{1}{5}$

$$2\frac{1}{5} \times 100\% = \frac{11}{\overset{}{5}} \times \overset{20}{\cancel{100}}\% = 220\%$$

   iii) 0.05

$$0.05 \times 100\% = 5\%$$

b) Write 36 as a percentage of 40.

First write '36 out of 40' as a fraction: $\dfrac{36}{40}$

Now change it to a percentage.

$\dfrac{\overset{9}{\cancel{36}}}{\cancel{40}} \times \cancel{100}\% = 90\%$

c) Write 15 cm as a percentage of 3 m.

First write '15 cm out of 3 m' as a fraction, remembering to change the units: $\dfrac{15\ \text{cm}}{300\ \text{cm}} = \dfrac{15}{300}$

Now change it to a percentage.

$\dfrac{\overset{5}{\cancel{15}}}{\cancel{300}} \times \cancel{100}\% = 5\%$

d) Write each of these percentages as a fraction.

   i) 45%

$$45\% = \frac{45}{100} = \frac{9}{20}$$

   ii) $2\frac{1}{2}\%$

$$2\frac{1}{2} = \frac{2\frac{1}{2}}{100} = \frac{5}{2} \times \frac{1}{100} = \frac{1}{40}$$

   iii) 140%

$$140\% = \frac{140}{100} = \frac{7}{5} = 1\frac{2}{5}$$

e) Write each of these percentages as a decimal.

   i) 25%

$$25\% = \frac{25}{100} = 0.25$$

   ii) 120%

$$120\% = \frac{120}{100} = 1.2$$

   iii) 0.25%

$$0.25\% = \frac{0.25}{100} = 0.0025$$

f) There are 60 books on a shelf. 35% of the books have hard covers.
How many soft-covered books are on the shelf?

Number of hard-covered books = 35% of 60

$$= \frac{35}{100} \times 60$$
$$= 21$$

Number of soft-covered books = 60 − 21 = 39
There are 39 soft-covered books.

g) Michael has 160 marbles.
Shaun has 40% more marbles than Michael.
How many marbles does Shaun have?

40% of 160 marbles $= \frac{40}{100} \times 160 = 64$ marbles

160 + 64 = 224
Shaun has 224 marbles.

h) 45% of the pupils in a class are boys.
There are 18 boys in the class.
How many pupils are in this class?
All of the pupils = 100%

45% of the pupils = 18

So 1% of the pupils $= \frac{18}{45}$

So 100% of the pupils $= \frac{18}{45} \times 100 = 40$

There are 40 pupils in this class.

i) Jill is 4% taller than she was one year ago.
Jill is now 135.2 cm tall.
How tall was she 1 year ago?

One year ago, Jill's height was 100%.
Now she is taller. Her new height is 100% + 4% = 104%
104% = 135.2 cm

So 1% $= \frac{135.2}{104}$ cm

So 100% $= \frac{135.2}{104} \times 100 = 130$ cm

One year ago, Jill's height was 130 cm.

D Simple financial calculations

Profit and loss

Anyone who sells something for **more** than they paid for it will make a profit. Anyone who sells something for **less** than they paid for it will make a loss.

The amount of money you pay to **buy** something is called the cost price. The amount of money you get when you **sell** something is called the selling price.

> Profit = selling price − cost price
> Loss = cost price − selling price

It is often useful to write the profit or loss as a **percentage**. The **percentage profit** or **percentage loss** is always calculated as a percentage of the **cost price**.

$$\text{Percentage profit} = \frac{\text{profit}}{\text{cost price}} \times 100\%$$

$$\text{Percentage loss} = \frac{\text{loss}}{\text{cost price}} \times 100\%$$

Examples

a) A music shop bought a CD for $8 and sold it for $10.
   i) How much profit did they make?

   Cost price (CP) = $8
   Selling price (SP) = $10
   Profit (P) = SP − CP
          = $10 − $8 = $2

   ii) What is the percentage profit?

   $$\text{Percentage profit} = \frac{\text{profit}}{\text{cost price}} \times 100\%$$
   $$= \tfrac{2}{8} \times 100\% = 25\%$$

b) In a sale, a shop sells a DVD player for $160.
   The shop paid $200 for it.
   i) How much was the shop's loss?

   Cost price (CP) = $200
   Selling price (SP) = $160
   Loss (L) = CP − SP
          = $200 − $160 = $40

ii) What is the percentage loss?

$$\text{Percentage loss} = \frac{\text{loss}}{\text{cost price}} \times 100\%$$
$$= \frac{40}{200} \times 100\%$$
$$= 20\%$$

Exercise 1

1 For each cost price and selling price, calculate:
   i) the profit or loss
   ii) the percentage profit or loss.

	Cost price	Selling price
a)	$4	$5
b)	$40	$45
c)	$60	$66
d)	$18	$19.80
e)	$20	$18
f)	$50	$48
g)	$72	$63
h)	$25	$21.50

2 A shopkeeper paid £40 for a suitcase and sold it for £65.

   a) How much profit did he make?
   b) What is the percentage profit?

3 In a sale, a shop sold a shirt for €19.50. The shop paid €20 for this shirt.

   a) How much was the shop's loss?
   b) What is the percentage loss?

4 Shop A bought a table for $50 and sold it for $70.
   Shop B bought the same kind of table for $45 and sold it for $65.
   Which shop made the biggest percentage profit?

5 A man buys 12 cameras for a total of $1800. He sells each camera
   for $186.
   What is his total percentage profit?

6 A florist buys 360 roses at $10 per dozen. She sells each rose for $1.10. What is her percentage profit?

7 Remi bought 60 dozen bananas at the market for $250. She sells them at 50¢ per banana. Before she can sell them all, 60 bananas go rotten and she must throw them away.
Calculate her percentage profit or loss.

Calculating cost price and selling price

If we know the cost price and the percentage profit/loss, we can calculate the selling price.

Example

A shopkeeper bought some shirts for £25 each and sold them at a loss of 15%.
Calculate the price the customers paid for these shirts.

Method 1
$$\text{Selling price} = \text{cost price} - \text{loss}$$
$$= £25 - (15\% \text{ of } £25)$$
$$= £25 - \left(\frac{15}{100} \times £25\right)$$
$$= £25 - £3.75$$
$$= £21.25$$

Method 2
$$\text{Selling price} = (100\% - 15\%) \text{ of } £25$$
$$= 85\% \text{ of } £25$$
$$= \frac{85}{100} \times £25$$
$$= £21.25$$

If we know the selling price and the percentage profit/loss, we can calculate the cost price.

Examples

a) A shop sells a jacket for $65 and makes 30% profit.
How much did the shop pay for the jacket (cost price)?

Method A (using proportion)
Cost price = 100%
$$\text{Selling price} = \text{cost price} + \text{profit}$$
$$= 100\% + 30\%$$
$$= 130\%$$
Selling price = $65

$$130\% = \$65$$

$$\text{So } 1\% = \frac{\$65}{130}$$

$$\text{So } 100\% = \frac{\$65}{130} \times 100$$

$$= \$50$$

Method B (using algebra)

Let the cost price $= \$x$

Profit $= 30\%$ of $\$x$

$$= \frac{3}{10} \times \$x$$

$$= \$\frac{3x}{10}$$

Profit $=$ selling price $-$ cost price

$$\frac{3x}{10} = 65 - x$$

$$3x = 650 - 10x$$

$$13x = 650$$

$$x = 50$$

The cost price of the jacket is $50.

b) John made a wooden cupboard and sold it for €24. He made a loss of 20%.

Calculate how much the cupboard had cost John to make.

Method A (using proportion)

Cost price $= 100\%$

Selling price $=$ cost price $-$ loss

$$= 100\% - 20\%$$

$$= 80\%$$

Selling price $=$ €24

$$80\% = €24$$

$$\text{So } 1\% = \frac{24}{€ \, 80}$$

$$\text{So } 100\% = \frac{24}{€ \, 80} \times 100$$

$$= €30$$

The cupboard cost John €30 to make.

Method B (using algebra)

Let the cost price $= €x$

Loss $= 20\%$ of $€x$

$$= \frac{20}{100} \times €x$$

$$= €\frac{2x}{10}$$

Loss $=$ cost price $-$ selling price

$$\frac{2x}{10} = x - 24$$

$$2x = 10x - 240$$

$$8x = 240$$

$$x = 30$$

The cupboard cost John €30 to make.

Exercise 2

1 Copy and complete this table.

Cost price	Selling price	Profit	% profit
$20			10%
	$30		20%
$145			5%

2 Copy and complete this table.

Cost price	Selling price	Loss	% loss
$40			10%
	$60.75		25%
$36			$12\frac{1}{2}\%$

3 A horse was bought for $2500. After winning a race, it was sold at a profit of 28%.
Calculate the selling price of this horse.

4 A furniture store bought a settee for $650. The store wants to make a profit of 16%.
Calculate the price you would pay for the settee in this store.

5 A musician buys a violin for $5000. After three years, she sells the violin and makes a $12\frac{1}{2}\%$ loss.
Calculate the selling price.

6 A man buys a bicycle for $240 from a shop. This shop makes a $33\frac{1}{3}\%$ profit.
Calculate how much the shop paid for the bicycle.

7 Mike buys a computer and sells it to Pete at a 25% profit. Pete makes a 20% loss when he sells it to Chris for $720.
How much did Mike pay for the computer?

8 Pam buys 300 CDs for a total of $1500. She sells 260 CDs at 20% above the cost price.
Pam sells the rest of the CDs at 50% loss.
Calculate Pam's total profit and percentage profit.

Discount

Sometimes, a shop will sell something at a price that is **less than** the normal selling price. Usually, a **percentage of the selling price** is subtracted. This amount is called a discount. The normal selling price is often called the marked price, as this is marked (written) on the price tag. When a discount is subtracted, the new, reduced, selling price is often called the sale price.

New selling price = old selling price − discount

or

Sale price = marked price − discount

Here are some of the reasons a shop may offer a discount on the price of an item:

- The item is damaged, dirty, or not quite as perfect as it should be.
- The item is no longer manufactured, so they want to sell the last ones they have.
- The shop has too many of an item.
- Last season's goods must be sold quickly to make space for the new season's goods (especially clothes).
- Loyal customers and shareholders of a business often get a discount.

Remember:

- Profit and loss are calculated as a percentage of the **cost price**.
- Discount is calculated as a percentage of the original **selling price**.
- The marked price is 100%, so the sale price is (100% − discount %) of marked price.

Examples

a) Sheila bought a dress with a discount of 15% on the marked price of $65.
How much did Sheila pay for the dress?

Discount = 15% of marked price

$$= \$\frac{15}{100} \times 65$$

$$= \$\frac{975}{100}$$

$$= \$9.75$$

Sale price = marked price − discount

$$= \$65 - \$9.75$$

$$= \$55.25$$

Sheila paid $55.25 for the dress.

b) A pair of shoes is marked at $175. To celebrate the shop's birthday, everything is on sale at a discount of 20%. A few weeks later the shop has a sale and discounts the shoes by an extra $17\frac{1}{2}$% of the sale price. At the end of the sale, the shoes are finally sold at a discount of 60% of the original marked price.

i) Calculate the first two sale prices of the shoes.

The marked price is 100%, so the 1st sale price will be $(100 - 20)$% of the marked price.

1st sale price = 80% of $175

$$= \$\frac{80}{100} \times 175$$

$$= \$140$$

2nd sale price = $(100 - 17\frac{1}{2})$% of 1st sale price

$$= 82\frac{1}{2}\% \text{ of 1st sale price}$$

$$= \$\frac{82.5}{100} \times 140$$

$$= \$115.50$$

ii) Calculate the final selling price of these shoes.

The final selling price is $(100 - 60)$% of the original marked price.

Final selling price = 40% of $175

$$= \$\frac{40}{100} \times 175$$

$$= \$70$$

c) Ead bought a table lamp in a sale with a discount of 10%. She paid $45 for the lamp.
What was the original marked price of this table lamp?

The discount is 10%, so the sale price is (100 − 10)% of the marked price.

90% of marked price = $45

$\frac{90}{100}$ × marked price = $45

marked price = $45 × $\frac{100}{90}$

= $50

The original marked price of the table lamp was $50.

Exercise 3

1 Copy and complete this table.

Marked price	Discount	% discount	Sale price
$100			$88
$30		10%	
$72		$12\frac{1}{2}$%	
		25%	$150
		12%	$77

2 The people who work at an electrical store are given $12\frac{1}{2}$% discount when they buy any items from the store. Andre buys a DVD recorder that is marked at $360.
How much does he pay?

3 A shopkeeper offers customers a discount if they buy many of the same items together.
One loaf of bread costs $1.20. If customers buy 10 loaves, they pay only $11.10.

a) What is the normal price of 10 separate loaves of bread?
b) What percentage discount does the shopkeeper give for buying 10 loaves?

4 Piet buys a TV in a sale and pays $390 after a 40% discount.
Calculate the original marked price.

5 A digital camcorder normally costs $2000. At the beginning of a sale, there is a discount of 20% on all items in the store. At the end of the sale, all items are discounted a further 30% of the first sale price.

a) Calculate the price of the camcorder at the beginning of the sale.
b) Calculate the price of the camcorder at the end of the sale.

Best buys

If you look in the shops, you will find many items to buy in different sizes, and with different prices. Sometimes it is important to know which one is the best value for money. To find this out, we must compare the prices for the different sizes, but we can only do this if we compare the prices for the **same amount** of the item.

Example

In Sam's supermarket, a 750 ml bottle of orange juice costs 45¢. A 2-litre bottle of the same orange juice costs $1.30. Which size is the better value for money?

We cannot compare 750 ml with 2 litres, so we must work out how much the **same amount** of each would cost.
There is no rule about how much this amount must be – we can choose any amount that is convenient.

Method 1
Let's compare the **same amount of orange juice** – we could choose 100 ml, or 10 ml, or 1 ml … and so on.
750 ml costs 45¢, so 10 ml will cost 45¢ ÷ 75 = 0.6¢.
2 litre (=2000 ml) costs $1.30, so 10 ml will cost $1.30 ÷ 200 = 0.65¢.
This means that if we buy the 2-litre bottle, we will pay more for 10 ml than if we buy the 750 ml bottle – or we can say that the smaller bottle costs less money for every 10 ml.

Method 2
Now let's compare the **same amount of money** – we could choose $1, 10¢, or 1¢, etc.
45¢ will buy 750 ml, so 10¢ will buy 750 ml ÷ 4.5 = 166.67 ml.
$1.30 will buy 2000 ml, so 10¢ will buy 2000 ml ÷ 13 = 153.85 ml.
This means that for our 10¢ we will get more orange juice out of the small bottle compared with the large bottle.
Both methods tell us that the smaller (750 ml) bottle of orange juice is better value for our money – and we call this the 'best buy'.

Exercise 4

1 Bob sells cans of olives in his store in two sizes. 460 g cans cost $1.30 and 700 g cans cost $3.96.
   Which size can of olives is better value for money?

2 Lisa can buy margarine at 58¢ for 250 g or $1.06 for 454 g.
   Which size is the better buy?

3 At the local market Balm can buy honey from Penny at 52¢ for a jar that weighs 454 g.
   Ann sells honey at 97¢ for a jar that weighs 822 g.
   Should Balm buy his honey from Ann or from Penny?

4 Abnash wants to buy a camcorder. Shop A and Shop B are both selling the same camcorder. At Shop A the price is $640 with a 30% discount. At shop B the camcorder costs $657 but has a special offer of $\frac{1}{3}$ off the price.
   Which shop will give Abnash the best buy?

5 Mohinda is looking for a washing machine. Shop A sells the one she wants for $480 and offers a 15% discount. Shop B sells the same machine at $468 with only $12\frac{1}{2}$% discount but also offers free washing powder for a year.
   Which shop offers the best buy?

Interest

When we save our money in a bank account, the bank will pay us some extra money because we allow them to use our money until we need it again. This extra money is called interest.

Sometimes we don't have all the money we need to pay for something. Then we can usually borrow money from a bank to help. When we pay back this money to the bank, we must pay some extra money because they allowed us to use their money for some time. This extra money is also called **interest**.

Interest is usually calculated as a **percentage** of the amount of money we save or borrow. This is called the interest rate – or sometimes just the **rate**.

Of course, if we save or borrow money for a longer time, the amount of interest should be more. So the interest rate will include an **amount of time** as well as the percentage. Most interest rates are given as a percentage for one year.

The Latin word for 'year' is 'annum' (that is why something that happens every year is called an 'annual' event), so

for one year = per year = per annum = p.a. (for short)

10% p.a. is an example of an interest rate.

Interest can be calculated for different periods of time – e.g. every month, or even every day!

The amount of money we save or borrow is called the principal or sometimes the capital.

Simple interest

Simple interest is always calculated using the **original principal**.

Simple interest = Principal × Rate × Time

or $$I = P \times \frac{R}{100} \times T$$

NOTE: The rate is always given as a percentage.
The time is always given in years.

Examples

a) A man saves $2000 for a year. His bank pays a simple interest rate of 6% p.a.

How much interest will he earn on his savings?

$$\text{Interest} = P \times \frac{R}{100} \times T$$

$$= \$2000 \times \frac{6}{100} \times 1$$

$$= \$120$$

The bank will pay him $120 interest.

b) Luca borrows $50 000 to start a business. He must pay a simple interest rate of 8% p.a.

What is the total amount that Luca must repay after four years?

First we must calculate the interest that Luca will pay for the four years.

$$\text{Interest} = P \times \frac{R}{100} \times T$$

$$= \$50 000 \times \frac{8}{100} \times 4$$

$$= \$16 000$$

At the end of the four years, Luca must repay the interest **and the principal**.

So at the end of the four years, Luca must repay
$16 000 + $50 000 = $66 000.

When someone borrows money, they normally repay part of it every month – they don't wait till the end of the time and pay it all. So the total amount to repay is calculated and divided over the number of months so that it is all repaid by the end of the total time.

In Example b) above, Luca has four years = 4 × 12 = 48 months to repay his loan.
So each month he must pay $66 000 ÷ 48 = $1375.

Examples

a) Simon needs $3400 for some new furniture. He already has $2500 and decides to save this with a bank to earn the rest of the money in interest. The bank pays 12% p.a. simple interest.
   How long will it be before Simon has all the money he needs?

   The simple interest that Simon's money must earn is
   $3400 − $2500 = $900.

   $$\text{Interest} = P \times \frac{R}{100} \times T$$

   $$900 = 2500 \times \frac{12}{100} \times T$$

   $$900 \times 100 = 2500 \times 12 \times T$$

   $$T = \frac{900 \times 100}{2500 \times 12} = 3$$

   Simon will have a total of $3400 after three years.

b) Bibi borrows some money for $2\frac{1}{2}$ years at 12% simple interest p.a. She repays an amount every month and $38 of this amount is interest.
   How much money did Bibi borrow?

   $2\frac{1}{2}$ years = $2\frac{1}{2}$ × 12 = 30 months
   Total interest paid is $38 × 30 = $1140

   $$\text{Interest} = P \times \frac{R}{100} \times T$$

   $$\$1140 = P \times \frac{12}{100} \times 2\frac{1}{2}$$

   $$\$1140 = P \times \frac{12}{100} \times \frac{5}{2}$$

   $$\$1140 \times 100 \times 2 = P \times 12 \times 5$$

   $$P = \frac{\$1140 \times 100 \times 2}{12 \times 5}$$

   $$= \$3800$$

   Bibi borrowed $3800.

c) Ian is buying a car. He borrows $20 000 for $3\frac{1}{2}$ years and pays $5880 in simple interest.

What annual rate of simple interest does Ian pay?

$$\text{Interest} = P \times \frac{R}{100} \times T$$

$$5880 = 20\,000 \times \frac{R}{100} \times 3\frac{1}{2}$$

$$5880 = 20\,000 \times \frac{R}{100} \times \frac{7}{2}$$

$$5880 \times 100 \times 2 = 20\,000 \times R \times 7$$

$$R = \frac{5880 \times 100 \times 2}{20\,000 \times 7}$$

$$= 8.4$$

Ian pays an interest rate of 8.4% p.a.

Exercise 5

1 Copy and complete this table.

Capital	Interest rate	Time	Simple interest	Total amount
$200	4%	3 years		
	10%	7 years	$700	
		4 years	$2800	$9800
	3%		$180	$2180
$40 000	$2\frac{1}{2}$%	5 years		
	$7\frac{1}{2}$%	3 years	$2250	
$10 000		4 years		$11 400
$6000	7%			$7680
$600	6%	$2\frac{1}{2}$ years		

2 A man saves $600 with a bank that pays simple interest of 7% p.a.
How long will it take before he has a total amount of $936?

3 A bank charges $55 simple interest on a personal loan for five months.
The interest rate at this bank is 12% p.a.
Calculate how much money the bank loaned.

4 Leo borrows $4000 and pays simple interest at $7\frac{1}{4}$% p.a. After two years, the interest rate is increased to 7.6% p.a.
Calculate how much Leo must repay at the end of seven years.

5 Jamie decides to buy a motorcycle. He cannot pay the full cash price of $1800, so he pays 18 equal monthly payments that include 7% p.a. simple interest.
How much does he pay each month?

Compound interest

Most banks do not calculate interest using the method of simple interest. Instead, they add together the principal and the interest at the end of each year to make a **new principal amount** and use this to calculate interest for the next year. This type of interest is called compound interest.

Often you will find that interest is compounded (calculated and added to the principal) each month; sometimes this is done every day!

Examples

a) Mr Lee deposits $10 000 in a bank. The bank pays 5% p.a. and the interest is compounded annually.
Calculate how much money Mr Lee has after two years.

1st year: Interest $= P \times \dfrac{R}{100} \times T$

$$= \$10\,000 \times \frac{5}{100} \times 1$$

$$= \$500$$

2nd year: **New** principal $=$ old principal $+$ interest

$$= \$10\,000 + \$500$$

$$= \$10\,500$$

Interest $= P \times \dfrac{R}{100} \times T$

$$= \$10\,500 \times \frac{5}{100} \times 1$$

$$= \$525$$

At the end of the 2nd year:

Total amount $=$ principal $+$ 1st year interest $+$ 2nd year interest

$$= \$10\,000 + \$500 + \$525$$

$$= \$11\,025$$

b) Calculate the interest paid on $2500 for one year at 4% p.a. if the interest is compounded every 6 months.

1st 6 months: Interest $= P \times \dfrac{R}{100} \times T$

$$= \$2500 \times \dfrac{4}{100} \times \dfrac{1}{2}$$

$$= \$50$$

2nd 6 months: New principal $= \$2500 + \$50 = \$2550$

$$\text{Interest} = \$2550 \times \dfrac{4}{100} \times \dfrac{1}{2}$$

$$= \$51$$

The total compound interest paid for this year is $50 + 51 = $101.

c) A man borrowed $50 000 to improve his house. The bank charged 6% p.a. interest, which was compounded every month.

i) Calculate the interest he paid at the end of the first month.

$$\text{Interest} = P \times \dfrac{R}{100} \times T$$

$$= \$50\,000 \times \dfrac{6}{100} \times \dfrac{1}{12}$$

$$= \$250$$

ii) At the end of the first month he paid $1000 to the bank. Calculate the interest he paid at the end of the second month.

At the end of the first month, the total amount owed is:
principal + interest − payment $= \$50\,000 + \$250 - \$1000$

$$= \$49\,250$$

The new principal for the 2nd month is $49 250.

$$\text{Interest} = P \times \dfrac{R}{100} \times T$$

$$= \$49\,250 \times \dfrac{6}{100} \times \dfrac{1}{12}$$

$$= \$246.25$$

iii) Calculate the total compound interest for the first two months.

Total compound interest $= \$250 + \$246.25$

$$= \$496.25$$

Calculating compound interest like this every month (or every day) would be a lot of work. Computers make it much quicker and easier these days, but the calculations they do are still the same.

Exercise 6

1 Copy and complete this table.

Capital	Interest rate	Time	Compound interest	Total amount
$5000	10%	2 years		
$20 000	5%	4 years		
$10 000	6%	3 years		
$5000	$11\frac{3}{4}$%	2 years		

2 Samir borrows $10 000 for three years at 2% p.a.

a) Calculate the simple interest for this period.
b) Calculate the compound interest if it is compounded each year.

3 Janet saves $800 with a bank that pays $12\frac{1}{2}$% p.a. interest compounded every six months.
Calculate how much money Janet has at the end of the first year.

4 Mira borrows $12 000 from a bank that charges 3% p.a. interest calculated monthly.
Calculate the total value of Mira's first three interest payments.

5 At the Savage Bank, interest is compounded every day.
Calculate the interest on $9000 at 2% p.a. after three days
(1 year = 365 days).

6 Sheila borrowed $90 000 to buy a flat. Interest of 4% p.a. is calculated at the end of every month. From the second month, Sheila repays $900 at the beginning of every month.
Calculate the total amount of compound interest Sheila repays in the first three months.

Hire purchase

We often need to buy expensive items when we don't have all the money to pay for them in full. Most businesses will be able to arrange for us to pay a part of the price now and the rest in smaller amounts every month for a period that is agreed until the full amount is paid.

We can describe this process in two parts:

- First we pay a small amount of money so that we can use the item for some time while it still belongs to the shop (in English we say that we '**hire**' the item).
- Then we pay money regularly until we own the item (in English we say that we '**purchase**' or 'buy' the item).

So this way of buying something is called hire purchase.

When we buy something in this way, we are really 'borrowing' the item from the business. We must pay some extra money to the business because they allowed us to use the item for some time before we had fully paid for it. This is very like what happens when we borrow money from a bank and, again, the extra money is called interest. It is usually calculated as **simple interest**.

- The amount of money we pay at the beginning is called the deposit (or down payment).
- The amount of money we pay each month is called an instalment.

Examples

A washing machine costs $460. Laura can pay a 10% deposit. She can pay the rest of the money in monthly instalments over two years with 8% simple interest.

a) Calculate the amount of Laura's monthly instalment.

Deposit: 10% of $\$460 = \dfrac{10}{100} \times \$460$

$= \$46$

Money still to pay: $\$460 - \$46 = \$414$

Interest: $I = P \times \dfrac{R}{100} \times T$

$= \$414 \times \dfrac{8}{100} \times 2$

$= \$66.24$

Total money to pay: $\$414 + \$66.24 = \$480.24$

Instalment: $480.24 is paid over two years.

Laura's monthly instalment is $\$480.24 \div 24 = \$20.01$.

b) How much more than the cash price will Laura pay for the washing machine?

Laura will pay deposit + 24 instalments $= \$46 + \$480.24$

$= \$526.24$

So Laura pays $\$526.24 - \$460 = \$66.24$ more than the cash price.

Exercise 7

1 Mr and Mrs Meyer bought a dining table and chairs. They paid a deposit of $125 and monthly instalments of $48 for $1\frac{1}{2}$ years.
Calculate the total amount they paid for the furniture.

2 Michael wants to buy a car but he has no money to pay a deposit. The car costs $50 000 and the simple interest will be 4% p.a. if he pays monthly instalments for five years.

a) Calculate the total amount Michael will pay for his car.
b) Calculate his monthly instalment.

3 An item costs $1200. To buy it on hire purchase, the deposit is $200 and the rest of the money is paid over $1\frac{1}{3}$ years at 15% p.a. simple interest.
Calculate the monthly instalment.

4 A computer package costs $1800. Sylvia pays a deposit of 15% and the rest of the money over three years at $9\frac{1}{2}$% p.a. simple interest.

a) Calculate Sylvia's monthly instalment.
b) How much more than the cash price does Sylvia pay?

5 The cash price of a sofa bed is $1600. To buy it on hire purchase you must pay a deposit of 25% and monthly instalments of $52 over $2\frac{1}{2}$ years.
Calculate the simple interest rate p.a. charged for this hire purchase.

Foreign currency and money exchange

Different countries around the world have different names for the money they use. The proper name for the money used in a country is currency.

The most common name for currency used in many countries is the **dollar**. Here are some of the countries that use the dollar as their currency. Can you think of any others?

Country	Currency name	Abbreviation
Australia	Australian dollar	A$
Canada	Canadian dollar	C$
Hong Kong	Hong Kong dollar	HK$
New Zealand	New Zealand dollar	NZ$
Singapore	Singapore dollar	S$
United States of America	US dollar	US$

Here are some of the other main currencies used around the world.

Country	Currency name	Abbreviation
United Kingdom	Pound sterling	£
Most countries in the European Union	Euro	€
China	Renminbi	人民币
India	Indian rupee	Rs
Japan	Japanese yen	¥

Find out the names of the currencies used in ten more countries around the world.

The currency used in one country has a different **value** from the currencies in other countries. It is even more complicated because the values of all the currencies around the world change every day!

If you need to know how to change one currency into another one, you can find the latest exchange rates listed in daily newspapers, at the main banks and on internet sites.

If you look at the list in a bank, you will notice that there are usually two exchange rates – one called '**buying**' and one called '**selling**'.

Here are some tips to help you decide whether to use the buying or selling rate:

- The exchange rates are given for the **banks** – so 'buying' means that the **bank** is buying, and 'selling' means that the **bank** is selling.
- The exchange rates are given for the banks to buy or sell foreign currency. So you must know which country you are in and what is the **home currency** (every other currency will be **foreign**). For example, the baht is the home currency in Thailand, but in America the baht will be foreign currency.

Examples

a) David is in New York, USA. He is travelling to South Africa and wants to exchange some US$ for South African Rand so that he has money for the taxi when he arrives.

  i) What is the foreign currency?

    David is in the USA, so the foreign currency is the Rand.

  ii) Is the bank buying or selling the foreign currency?

    The bank will give David Rand and keep his US$, so the bank is selling the foreign currency.

b) David has arrived in Johannesburg, South Africa. He goes to the bank to change some more of his US$ for Rand.

   i) What is the foreign currency?

   David is in South Africa, so the foreign currency is US$.

   ii) Is the bank buying or selling the foreign currency?

   The bank will give David Rand and keep his US$, so the bank is buying the foreign currency.

The list of exchange rates you find at a bank usually tells you how much of the **home currency** you will get for **one unit of the foreign currency**.

The value of currencies changes all the time, and the exchange rates are different in different countries. Often the exchange rates are different at different banks in the same country!

The simplest way to find the value of any amount of one currency in another currency is to use one of the many currency converter websites. These websites will always use the latest exchange rates and you don't have to worry about who is buying and who is selling.

When you know the right exchange rate for the currencies you are working with, the actual conversion calculations are really quite straight-forward proportion.

Examples

These exchange rates were correct (to 5 s.f.) in July 2008. Use them to work out the answers to the following questions.

€1.0 = US$1.5899

US$1.0 = €0.628 97

a) Heinrich lives in Stuttgart, Germany. He is going to study philosophy at the University of San Francisco, USA. His student fees will cost US$3675 for a year. How many euros must Heinrich borrow to pay for his studies?

We know that US$1 = €0.628 97
So US$3675 = €0.628 97 × 3675
          = €2311.4647
Heinrich will need to borrow €2311.46 (to the nearest ¢).

b) Brad and Janet live in Arkansas, USA. They are planning their honeymoon in Florence, Italy. The hotel they want to stay at costs €78.50 per night.

How much will the hotel cost in US$ if they stay in Florence for a week?

One night costs €78.50
So seven nights cost 7 × €78.50 = €549.50
We know that €1 = US$1.5899
So €549.50 = US$ 1.5899 × 549.5
　　　　　　= US$873.650 05
Brad and Janet's honeymoon hotel will cost US$873.65 (to the nearest ¢).

Activity

Imagine you are planning a holiday in a foreign country.

1   Choose the country you would like to visit. Then choose one city to visit in this country, and decide how long you will stay.
2   Find out how much of your country's money it will cost you to travel to this city.
3   a)   Find out the cost of a hotel in this city for one night.
　　 b)   Make a list of all the things you think you will need to buy while you are staying in this city. Now find out how much these things cost in the currency of the country you want to visit.
4   Find out the exchange rate between this country's currency and the currency you use in your own country, then convert all the prices you have found into the currency you use in your country.
5   Work out how much money (in your country's currency) you will need to pay for your holiday. Do you think that it is cheap or expensive to live in this country compared with your own country?

NOTE: You can get a lot of this information from the internet. If you do not have access to the internet, your teacher will be able to give you a list of the prices of some common items in a few countries, and the exchange rate between their currency and yours.

Square roots and cube roots

Key vocabulary

absolute value	perfect cube	square root
cube root	perfect square	squared
cubed	prime factor	

A Square roots

If a number is multiplied by itself, we say that the number is squared, and the answer is the **square** of that number.

For example:

$5 \times 5 = 5^2 = 25$ 25 is the square of 5 (written as 5^2).

$3.2 \times 3.2 = (3.2)^2 = 10.24$ 10.24 is the square of 3.2 [written as $(3.2)^2$].

The square of a **whole number** (an **integer**) will also be a whole number (an integer).
These squares are called perfect squares.

25 is a perfect square, but 10.24 is not a perfect square.

5^2 can be read as:

- '5 squared'
- 'the square of 5'
- '5 to the power of 2'.

Suppose we want to know which number, when squared, gives 25. From the example above, we know that the answer is 5. We call this number the square root – so 5 is the square root of 25.

A 'root' is the place where a plant 'starts' or 'comes from'. In the same way, a 'square root' is where a 'square' comes from – or what we use to make one.

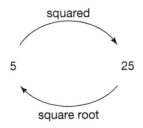

We use the symbol $\sqrt{}$ for a square root. An **index of** $\frac{1}{2}$ also represents a square root.

$$\sqrt{25} = (25)^{\frac{1}{2}} = \sqrt{5 \times 5} = \sqrt{5^2} = 5$$

+ or − sign?

We know that $5 \times 5 = 25$.
$(-5) \times (-5) = 25$ as well.
So, to be accurate, if $x^2 = 25$

$$x = \pm\sqrt{25}$$
$$x = \pm 5 \text{ (read as '}+5 \text{ or } -5\text{')}$$

NOTE: The $\pm$ sign means $+$ *or* $-$.

Usually we take the positive answer (root) unless there is a special reason to take the negative answer.

This positive root is called the absolute value of the square root and can be written with two vertical lines: $|\sqrt{25}| = +5$

Perfect squares

We know that the square of each **whole number** is called a perfect square, because it is also a whole number.

You should try to remember the squares of all the numbers between 1 and 25. This will help you to recognise them in calculations, and make things easier.

$1^2 = 1$	$6^2 = 36$	$11^2 = 121$	$16^2 = 256$	$21^2 = 441$
$2^2 = 4$	$7^2 = 49$	$12^2 = 144$	$17^2 = 289$	$22^2 = 484$
$3^2 = 9$	$8^2 = 64$	$13^2 = 169$	$18^2 = 324$	$23^2 = 529$
$4^2 = 16$	$9^2 = 81$	$14^2 = 196$	$19^2 = 361$	$24^2 = 576$
$5^2 = 25$	$10^2 = 100$	$15^2 = 225$	$20^2 = 400$	$25^2 = 625$

NOTE: Remember, if you know that $4^2 = 16$, then you also know that $\sqrt{16} = 4$, and so on.

Revision

Factors

We have already learned that a whole number can be written as the product of two or more other whole numbers.

For example:

$12 = 1 \times 12 \qquad 12 = 2 \times 6 \qquad 12 = 3 \times 4$

So, the **factors** of 12 are 1, 2, 3, 4, 6 and 12.

Whole numbers that multiply to give a certain number are called **factors** of that number.

We can use factors and the list of squares on page 104 to work out many other squares more easily.

For example:

28^2 can be written using factors as $(2 \times 14)^2 = 2^2 \times 14^2$
$$= 4 \times 196$$
$$= 784$$

So $28^2 = 784$

Activity

Using factors and the list of squares, work out these perfect squares.

a) 30^2 b) 50^2
c) 120^2 d) 250^2

Revision

Prime factors

Some numbers have only **two factors** – the number itself and 1. These numbers are called **prime numbers**.

NOTE: 1 is **not** a prime number as it has only one factor.

Prime factors are those factors of a number that are **prime numbers**.
We have already learned that every whole number can be written as the product of its prime factors.

For example:
The prime factors of 12 are 2 and 3.
We can write 12 as $2 \times 2 \times 3$ or $2^2 \times 3$.

Finding square roots using prime factors

Follow these steps to find the square root of a number.

1 Start by dividing the number by its lowest prime factor.
2 Divide the remainder by the same lowest prime factor again (if you can) until it cannot divide exactly any more.
3 Now use the next-lowest prime factor to divide the remainder of the number until it cannot divide exactly any more.
4 Continue like this until the **remainder** is also a **prime factor**.
5 Write the number as the product of all these prime factors.
6 Work out the square root from these prime factors.

NOTE: The first five steps above are exactly the same as the method we learned in Coursebook 1 Unit 2 to find all the prime factors of a number.

Examples

a) Find all the prime factors of 81 and use them to find the square root of 81.

81 will not divide by 2, so we try the next prime number, 3.
(81 will divide by 3.)

$3\overline{)81}$
27 (remainder)

27 will also divide by 3.

$3\overline{)27}$
9

And so will 9.

$3\overline{)9}$
3

The remainder 3 is also a prime factor.
Seen altogether, our division process is:

$3\overline{)81}$
$3\overline{)27}$
$3\overline{)\,9}$
3

As a product of all its prime factors, $81 = 3 \times 3 \times 3 \times 3$
$= (3 \times 3) \times (3 \times 3)$
$= (3 \times 3)^2$
$= 9^2$

So $\sqrt{81} = \sqrt{9^2}$
$= 9$

b) Find all the prime factors of 225 and use them to work out the square root of 225.

225 will not divide by 2, but it will divide by 3.
3)225
3) 75   (75 will divide by 3)
5) 25   (25 will not divide by 3, but it will divide by 5)
    5
So 225 = 3 × 3 × 5 × 5
       = (3 × 5) × (3 × 5)
       = (3 × 5)²
       = (15)²
So $\sqrt{225} = \sqrt{(15)^2}$
       = 15

We can use the **prime factor method** to calculate the square root of a number, even if it is not a perfect square.

Example

Find the square root of 48, using the prime factor method.

2)48
2)24
2)12
2) 6
   3
So 48 = 2 × 2 × 2 × 2 × 3
      = (2 × 2) × (2 × 2) × 3
      = (2 × 2)² × 3
      = 4² × 3
So $\sqrt{48} = \sqrt{4^2 \times 3}$
      $= \sqrt{4^2} \times \sqrt{3}$
      $= 4 \times \sqrt{3}$
      = 4 × 1.732
      = 6.928

NOTE: Try to memorise the square roots of the first few prime numbers.

$\sqrt{2} \approx 1.414$ $\sqrt{3} \approx 1.732$ $\sqrt{5} \approx 2.236$ $\sqrt{7} \approx 2.646$ $\sqrt{11} \approx 3.317$
(Remember, these are only approximations of endless decimals.)

 **Exercise 1**

1 a) Write down all the numbers smaller than 100 that are perfect squares.
   b) Write down all the numbers between 200 and 400 that are perfect squares.

2 Write down the value of each of these.
   a) $\sqrt{25}$ b) $(100)^{\frac{1}{2}}$ c) $\sqrt{64}$ d) $(49)^{\frac{1}{2}}$

3 Use the prime factor method to calculate the square root of each of these numbers.
   a) 289   b) 441   c) 529   d) 324
   e) 484   f) 1936   g) 3136   h) 4096
   i) 4900   j) 8649   k) 7056   l) 1024

4 $41^2 = 1681$. Use this to find the square root of 168 100.

5 $\sqrt{841} = 29$. Use this to find the value of $\sqrt{3364}$.

6 $31^2 = 961$. Use this to find the square of 310.

7 $\sqrt{1369} = 37$. Use this to find the value of 370^2.

8 a) Which of these numbers is bigger?
      $\sqrt{144} + \sqrt{25}$ *or* $\sqrt{144 + 25}$
   b) How much bigger is it?

9 a) What is the value of $\sqrt{625}$?
   b) Use your answer to work out $\sqrt{6.25}$.

10 Use the method you used in question **9** to work out these square roots.
   a) $\sqrt{0.25}$ b) $(5.29)^{\frac{1}{2}}$ c) $\sqrt{0.0256}$ d) $(1.69)^{\frac{1}{2}}$

Calculating square roots using trial and improvement

We can use the values of the perfect squares that we know (or can find in tables) to find two numbers so that the square root of another number (which we want to find) is between these numbers. We then use trial and improvement to reach an estimated value of the required degree of accuracy.

Example

Work out an estimated value for the square root of 42.6.

We know that $6^2 = 36$ and $7^2 = 49$.

Now, 42.6 is between 36 and 49, so $\sqrt{42.6}$ must be between 6 and 7. So we start with a number halfway between them, 6.5.

$6.5 \times 6.5 = 42.25$ This is still less than 42.6, so we try a number slightly bigger.

$6.6 \times 6.6 = 43.56$ This is now too big, so we try a number between 6.5 and 6.6.

$6.55 \times 6.55 = 42.9025$ This is still too big, so we try a slightly smaller number.

$6.54 \times 6.54 = 42.7716$ This is closer, but still too big, so we try a slightly smaller number.

$6.53 \times 6.53 = 42.6409$ This is now very close.

We must decide what level of accuracy we need our answer to have. We are usually told to estimate the square root to a given number of decimal places accuracy.
We now have an answer that is between 6.5 and 6.53 – so the square root of 42.6 is 6.5 correct to 1 decimal place.

NOTE: Do at least one trial to **one more decimal place** than you need for the accuracy, so that you can round your answer.

In the example above, we could keep working and improve our level of accuracy.

$6.52 \times 6.52 = 42.5104$ This is still less than 42.6, so we try a number between 6.53 and 6.52.

$6.525 \times 6.525 = 42.575625$ This is still too small, so we try 6.526.

$6.526 \times 6.526 = 42.588676$ This is also too small, so we try 6.527.

$6.527 \times 6.527 = 42.601729$ This is very close indeed.

We can continue this method of trial and improvement until we find an answer as accurate as we need.
So far, we can say that the square root of 42.6 lies between 6.526 and 6.527.
This means that $\sqrt{42.6} = 6.53$ correct to 2 decimal places.

This method takes some time, but we can use it to calculate a very accurate estimate of the square root of any number.

Sometimes we want to calculate the square root of a number that is not between two perfect squares that we know. We can then use a table of squares and square roots like the one below to find these two numbers to start the trial and improvement calculation process.

Square root	Number	Square	Square root	Number	Square	Square root	Number	Square
1.000	1	1	5.916	35	1225	8.307	69	4761
1.414	2	4	6.000	36	1296	8.367	70	4900
1.732	3	9	6.083	37	1369	8.426	71	5041
2.000	4	16	6.164	38	1444	8.485	72	5184
2.236	5	25	6.245	39	1521	8.544	73	5329
2.449	6	36	6.325	40	1600	8.602	74	5476
2.646	7	49	6.403	41	1681	8.660	75	5625
2.828	8	64	6.481	42	1764	8.718	76	5776
3.000	9	81	6.557	43	1849	8.775	77	5929
3.162	10	100	6.633	44	1936	8.832	78	6084
3.317	11	121	6.708	45	2025	8.888	79	6241
3.464	12	144	6.782	46	2116	8.944	80	6400
3.606	13	169	6.856	47	2209	9.000	81	6561
3.742	14	196	6.928	48	2304	9.055	82	6724
3.873	15	225	7.000	49	2401	9.110	83	6889
4.000	16	256	7.071	50	2500	9.165	84	7056
4.123	17	289	7.141	51	2601	9.220	85	7225
4.243	18	324	7.211	52	2704	9.274	86	7396
4.359	19	361	7.280	53	2809	9.327	87	7569
4.472	20	400	7.348	54	2916	9.381	88	7744
4.583	21	441	7.416	55	3025	9.434	89	7921
4.690	22	484	7.483	56	3136	9.487	90	8100
4.796	23	529	7.550	57	3249	9.539	91	8281
4.899	24	576	7.616	58	3364	9.592	92	8464
5.000	25	625	7.681	59	3481	9.644	93	8649
5.099	26	676	7.746	60	3600	9.695	94	8836
5.196	27	729	7.810	61	3721	9.747	95	9025
5.292	28	784	7.874	62	3844	9.798	96	9216
5.385	29	841	7.937	63	3969	9.849	97	9409
5.477	30	900	8.000	64	4096	9.899	98	9604
5.568	31	961	8.062	65	4225	9.950	99	9801
5.657	32	1024	8.124	66	4356	10.000	100	10000
5.745	33	1089	8.185	67	4489			
5.831	34	1156	8.246	68	4624			

Example

Estimate the square root of 1063 using trial and improvement.

Look at the table on page 110 to find the two perfect squares that are closest to 1063 – look in the 'Square' column to find a number that is just smaller than 1063, and one that is just bigger.
We see from the table that $32^2 = 1024$ and $33^2 = 1089$.
This means that $\sqrt{1063}$ must be somewhere between 32 and 33.
So we start with a number halfway between them, 32.5.

$32.5 \times 32.5 = 1056.25$ This is too small, so we try 32.6.
$32.6 \times 32.6 = 1062.76$ This is very close, so let's try 32.61.
$32.61 \times 32.61 = 1063.4121$

We can say that the square root of 1063 lies between 32.6 and 32.61.
So an answer of 32.6 would be accurate to 1 decimal place.
To improve the accuracy let's try 32.605 next.
$32.605 \times 32.605 = 1063.086025$
The answer is now between 32.60 and 32.605.
This means that $\sqrt{1063} = 32.60$ correct to 2 decimal places.

Using a calculator to find the square root of any number

The $\sqrt{}$ key can be found on all modern calculators. We can use it to find the square root of any number.

Exercise 2

1 For each of these square roots, find one perfect square (whole number) that is just bigger than the square root, and one that is just smaller.
 a) $\sqrt{45}$ b) $\sqrt{30}$ c) $\sqrt{72.8}$ d) $\sqrt{115}$
 e) $\sqrt{423.3}$ f) $\sqrt{500}$

2 Use the method of trial and improvement to estimate the square root of each number in question **1**. Give your answers correct to 2 decimal places.

3 Use trial and improvement to estimate the square root of each of these numbers. Use the table on page 110 to help you, and give your answers correct to 2 decimal places.
 a) $\sqrt{811}$ b) $\sqrt{1397}$ c) $\sqrt{4145.7}$ d) $\sqrt{6613.9}$

B Cube roots

If a number is multiplied by itself, and the product is multiplied by the first number again, we say that the number is cubed, and the answer is the **cube** of the number.

For example:
$3 \times 3 \times 3 = 9 \times 3 = 27 = 3^3$

The cube of a whole number (an integer) will also be a whole number (an integer).
These cubes are called perfect cubes.

3^3 is a perfect cube.

3^3 can be read as:

- '3 cubed'
- 'the cube of 3'
- '3 to the power of 3'.

Suppose we want to know which number, when cubed, gives 27. From the example above, we know that the answer is 3. We call this number the cube root – so 3 is the cube root of 27.
A 'root' is the place where a plant 'starts' or 'comes from'. In the same way, a 'cube root' is where a 'cube' comes from – or what we use to make one.

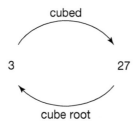

We use the symbol $\sqrt[3]{\ }$ for a cube root. An **index of** $\frac{1}{3}$ also represents a cube root.

$$\sqrt[3]{27} = (27)^{\frac{1}{3}} = \sqrt[3]{3 \times 3 \times 3} = \sqrt[3]{3^3} = 3$$

We know that $3 \times 3 \times 3 = 27$.
But $(-3) \times (-3) \times (-3) = (-27)$, or $\sqrt[3]{(-27)} = -3$.
So it is possible to find the **cube root** of a **negative** number. This is different from squares and square roots – the square root of a negative number does not exist.

Try to memorise the first 10 perfect cubes so you can recognise them when you do calculations:

$$1^3 = 1 \qquad\qquad 6^3 = 216$$
$$2^3 = 8 \qquad\qquad 7^3 = 343$$
$$3^3 = 27 \qquad\qquad 8^3 = 512$$
$$4^3 = 64 \qquad\qquad 9^3 = 729$$
$$5^3 = 125 \qquad\qquad 10^3 = 1000$$

Finding cube roots using prime factors

We can calculate the cube root of a number using its prime factors, using the same method as we used to calculate the square root.

Example

Find all the prime factors of 9261 and use them to calculate the cube root of 9261.

First, we find all the prime factors.
9261 will divide by 3.

$$\begin{array}{r} 3\underline{)9261} \\ 3\underline{)3087} \\ 3\underline{)1029} \\ 7\underline{)\ 343} \\ 7\underline{)\ \ 49} \\ 7 \end{array}$$ (343 won't divide by 3 or 5, so we try the next prime number, 7.)

So $9261 = 3 \times 3 \times 3 \times 7 \times 7 \times 7$
$$= 3^3 \times 7^3$$

So $\sqrt[3]{9261} = \sqrt[3]{3^3 \times 7^3}$
$$= 3 \times 7$$
$$= 21$$

Calculating cube roots using trial and improvement

We can estimate the cube root of a number that is not a perfect cube using trial and improvement, just as we did for square roots.

Example

Use the method of trial and improvement to estimate $\sqrt[3]{18.6}$, accurate to 1 decimal place.

We know that $2^3 = 8$ and $3^3 = 27$, so the cube root of 18.6 must be between 2 and 3.

So we start with 2.5.

$2.5 \times 2.5 \times 2.5 = 15.625$ Too small, try 2.6.
$2.6 \times 2.6 \times 2.6 = 17.576$ Too small, try 2.7.
$2.7 \times 2.7 \times 2.7 = 19.683$ Too big, try 2.65 (between 2.6 and 2.7).
$2.65 \times 2.65 \times 2.65 = 18.609\,625$

This shows that the cube root of 18.6 lies between 2.6 and 2.65.
So $\sqrt[3]{18.6} = 2.6$ correct to 1 decimal place.

NOTE: Remember to do at least one trial to **one more decimal place** than you need for the accuracy, so that you can round your answer.

Exercise 3

1 Write down all the perfect cubes that are less than 1000.

2 Write down the value of each of these.
   a) $\sqrt[3]{125}$ b) $(1000)^{\frac{1}{3}}$ c) $\sqrt[3]{343}$ d) $(729)^{\frac{1}{3}}$

3 Use the prime factor method to calculate the cube root of each of these numbers.
   a) 1728   b) 2197   c) 2744   d) 3375
   e) 4096   f) 8000

4 $\sqrt[3]{4913} = 17$. Use this to find the value of 170^3.

5 a) What is $\sqrt[3]{125}$?
   b) Use your answer to work out $\sqrt[3]{0.125}$.

6 Use the method you used in question **5** to work out each of these cube roots.
   a) $\sqrt[3]{0.064}$ b) $(0.512)^{\frac{1}{3}}$ c) $\sqrt[3]{0.000216}$

7 Use trial and improvement to estimate each of these cube roots.
   Give your answers accurate to 1 decimal place.
   a) $\sqrt[3]{42.7}$ b) $\sqrt[3]{99.9}$ c) $\sqrt[3]{157.3}$ d) $\sqrt[3]{589.4}$

Some properties of square roots

These rules are true for all square roots:

Rule 1 $\sqrt{ab} = \sqrt{a} \times \sqrt{b}$ where a and $b \geq 0$.

Rule 2 $\sqrt{a}(m + n) = \sqrt{a}(m) + \sqrt{a}(n)$ where $a \geq 0$ and m and n are any rational numbers.

Rule 3 $\sqrt{\dfrac{a}{b}} = \dfrac{\sqrt{a}}{\sqrt{b}}$ where $a \geq 0$ and $b > 0$.

NOTE: When you are simplifying square roots, look at the number inside the $\sqrt{}$. Are any of its **factors** perfect squares?

Examples

a) Write each of these in its simplest square root form.

   i) $\sqrt{18}$

$$\begin{aligned} \sqrt{18} &= \sqrt{9 \times 2} \\ &= \sqrt{9} \times \sqrt{2} \quad \text{(Using rule 1)} \\ &= \sqrt{3^2} \times \sqrt{2} = 3\sqrt{2} \end{aligned}$$

   ii) $\sqrt{80}$

$$\begin{aligned} \sqrt{80} &= \sqrt{16 \times 5} \\ &= \sqrt{16} \times \sqrt{5} \quad \text{(Using rule 1)} \\ &= \sqrt{4^2} \times \sqrt{5} = 4\sqrt{5} \end{aligned}$$

b) Given that $\sqrt{2} \approx 1.414$, simplify $\sqrt{50}$ and give your answer correct to 2 decimal places.

$$\begin{aligned} \sqrt{50} &= \sqrt{25 \times 2} \\ &= \sqrt{25} \times \sqrt{2} \quad \text{(Using rule 1)} \\ &= \sqrt{5^2} \times \sqrt{2} \\ &= 5\sqrt{2} \\ &= 5 \times 1.414 = 7.07 \end{aligned}$$

c) Simplify $\sqrt{2}(\sqrt{2} + \sqrt{8})$.

Using rule 2

$$\begin{aligned} \sqrt{2}(\sqrt{2} + \sqrt{8}) &= \sqrt{2} \times \sqrt{2} + \sqrt{2} \times \sqrt{8} \\ &= \sqrt{4} + \sqrt{16} \quad \text{(Using rule 1)} \\ &= \sqrt{2^2} + \sqrt{4^2} \\ &= 2 + 4 = 6 \end{aligned}$$

d) Simplify $\sqrt{\dfrac{72}{20}}$ if $\sqrt{2} \approx 1.414$ and $\sqrt{5} \approx 2.236$. Give your answer correct to 2 decimal places.

$$\sqrt{\dfrac{72}{20}} = \dfrac{\sqrt{72}}{\sqrt{20}} \qquad \text{(Using rule 3)}$$

$$= \dfrac{\sqrt{36 \times 2}}{\sqrt{4 \times 5}}$$

$$= \dfrac{\sqrt{36} \times \sqrt{2}}{\sqrt{4} \times \sqrt{5}} \qquad \text{(Using rule 1)}$$

$$= \dfrac{6\sqrt{2}}{2\sqrt{5}}$$

$$= \dfrac{3\sqrt{2}}{\sqrt{5}}$$

$$= \dfrac{3 \times 1.414}{2.236}$$

$$= 1.90 \text{ (correct to 2 decimal places)}$$

e) Simplify $\sqrt{108} - \sqrt{48} + \sqrt{75}$ and give your answer correct to 2 decimal places.
You are given that $\sqrt{3} \approx 1.732$.

$$\sqrt{108} - \sqrt{48} + \sqrt{75} = \sqrt{2^2 \times 3^2 \times 3} - \sqrt{2^2 \times 2^2 \times 3} + \sqrt{5^2 \times 3}$$

$$= (\sqrt{2^2} \times \sqrt{3^2} \times \sqrt{3}) - (\sqrt{2^2} \times \sqrt{2^2} \times \sqrt{3})$$

$$+ (\sqrt{5^2} \times \sqrt{3}) \qquad \text{(Using rule 1)}$$

$$= (2 \times 3 \times \sqrt{3}) - (2 \times 2 \times \sqrt{3}) + (5 \times \sqrt{3})$$

$$= 6\sqrt{3} - 4\sqrt{3} + 5\sqrt{3}$$

$$= \sqrt{3}(6 - 4 + 5) \qquad \text{(Using rule 2)}$$

$$= 7\sqrt{3}$$

$$= 7 \times 1.732$$

$$= 12.12 \text{ (correct to 2 decimal places)}$$

f) Remove the brackets and simplify $\sqrt{2}(\sqrt{6} + \sqrt{2})$.

$$\sqrt{2}(\sqrt{6} + \sqrt{2}) = (\sqrt{2} \times \sqrt{6}) + (\sqrt{2} \times \sqrt{2}) \qquad \text{(Using rule 2)}$$

$$= (\sqrt{2} \times \sqrt{2 \times 3}) + 2$$

$$= (\sqrt{2} \times \sqrt{2} \times \sqrt{3}) + 2 \qquad \text{(Using rule 1)}$$

$$= 2\sqrt{3} + 2$$

$$= (2 \times 1.732) + 2 \quad (\sqrt{3} \approx 1.732)$$

$$= 3.464 + 2$$

$$= 5.464$$

Exercise 4

1 Simplify each of these square roots. Write your answers using square roots of the smallest numbers possible.

a) $\sqrt{12}$ b) $\sqrt{27}$ c) $\sqrt{45}$ d) $\sqrt{32}$

e) $\sqrt{54}$ f) $\sqrt{24}$ g) $\sqrt{98}$ h) $\sqrt{63}$

i) $\sqrt{200}$ j) $\sqrt{128}$ k) $\sqrt{112}$ l) $\sqrt{175}$

2 Simplify each of these square roots. Write your answers using square roots of the smallest numbers possible.

a) $\sqrt{\frac{9}{4}}$ b) $\sqrt{\frac{25}{16}}$ c) $\sqrt{\frac{18}{8}}$ d) $\sqrt{\frac{24}{9}}$

e) $\sqrt{\frac{6}{4}}$ f) $\sqrt{\frac{12}{15}}$

3 Simplify each of these. Write your answers using square roots of the smallest numbers possible.

a) $2\sqrt{5} \times 3\sqrt{5}$ b) $\sqrt{12} \times 2\sqrt{3}$ c) $2\sqrt{6} \times \sqrt{3}$

d) $(\sqrt{5})^2 \times (-3\sqrt{7})$ e) $(-10\sqrt{13}) \times \left(-4\sqrt{\frac{1}{13}}\right)$ f) $\left(\sqrt{8} \times \sqrt{\frac{1}{2}}\right) \times 5\sqrt{\frac{1}{2}}$

4 Calculate each of these correct to 2 decimal places. Use $\sqrt{2} \approx 1.414$ and $\sqrt{3} \approx 1.732$.

a) $\sqrt{12}$ b) $\sqrt{150}$ c) $-\sqrt{216}$ d) $2\sqrt{3} - 3\sqrt{2}$

5 Calculate each of these correct to 3 decimal places. Use $\sqrt{2} \approx 1.414$, $\sqrt{3} \approx 1.732$ and $\sqrt{7} \approx 2.646$.

a) $4\sqrt{2} + 3\sqrt{2} - 7\sqrt{2}$ b) $\sqrt{128} - \sqrt{98} + \sqrt{72}$

c) $2\sqrt{8} - \sqrt{18} + \sqrt{50}$ d) $\sqrt{6} + \sqrt{24} - \sqrt{7} - \sqrt{28}$

6 Simplify each of these calculations as much as possible.

a) $2 \times \sqrt{3} \times \sqrt{3}$ b) $2\sqrt{2}(2\sqrt{8} + \sqrt{2})$

c) $\sqrt{5}(\sqrt{10} - \sqrt{5})$ d) $\frac{1}{3\sqrt{3}}(4\sqrt{3} - \sqrt{3})$

e) $2\sqrt{3}(\sqrt{27} - \sqrt{12})$ f) $(7\sqrt{5} - \sqrt{5} + \sqrt{180})\frac{3}{4\sqrt{5}}$

Unit 6 Real numbers

Key vocabulary

imaginary number	rationalise	terminating decimal
irrational number	real number	
rational number	recurring decimal	

 ## Real numbers

All the numbers that you will work with in mathematics at school level are part of a group of numbers called real numbers.

Every type of number is part of a group with its own special properties. All these groups of numbers are related as shown in the family tree of numbers below:

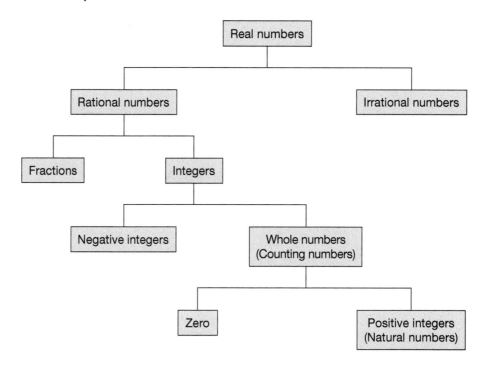

Rational numbers

Any number that can be written as an **exact fraction** is called a rational number.

> So **any** rational number can be written as $\dfrac{a}{b}$ where a is an integer, b is an integer and $b \neq 0$.

Revision

Fractions

- The **numerator** in a fraction is the top number.
- The **denominator** in a fraction is the bottom number.
- A **proper fraction** is one where the numerator is smaller than the denominator.
  For example: $\frac{1}{7}$ $\frac{2}{3}$ $\frac{3}{4}$
- An **improper fraction** is one where the numerator is bigger than or equal to the denominator.
  For example: $\frac{4}{3}$ $\frac{3}{2}$ $\frac{7}{7}$
- A **mixed number** has an integer part and a proper fraction part.
  For example: $1\frac{1}{2}$ $2\frac{2}{3}$ $3\frac{1}{7}$

Some properties of rational numbers

1 If x and y are any two rational numbers, then
  $x + y = z$, where z is another rational number.
2 If x and y are any two rational numbers, then $x + y = y + x$.
3 If x, y and z are any three rational numbers, then
  $x + (y + z) = (x + y) + z$.
4 If x is any rational number, then $x + 0 = 0 + x = x$.
5 If x and y are any two rational numbers, then
  $x - y = x + (-y)$ and $x + (-y) = z$, where z is another rational number (property 1).
6 If x and y are any two rational numbers, then
  $x - y$ must also be a rational number.
7 If x is a rational number, then $x - 0 = x$.
8 If x and y are any two rational numbers, then
  $x \times y = z$, where z is another rational number.
9 If x and y are any two rational numbers, then $x \times y = y \times x$.
10 If x, y and z are any three rational numbers, then
  $x \times (y \times z) = (x \times y) \times z$.
11 If x is any rational number, then $x \times 1 = x$.

12 If x is any rational number, then $x \times 0 = 0$.

13 If x, y and z are any three rational numbers, then
$x \times (y + z) = (x \times y) + (x \times z)$.

14 If x and y are two rational numbers and both x and $y \neq 0$, $\dfrac{x}{y}$ is a rational number, and the reciprocal, $\dfrac{y}{x}$, is also a rational number.

15 If x and y are two rational numbers, then
$$x \div y = x \times \frac{1}{y}, \text{ where } y \neq 0.$$

16 If x and y are any two rational numbers and $x < y$, then
$$x < \frac{x + y}{2} < y.$$

Fractions and decimals

Revision

We have already learned that:

- All fractions can be written as decimals.
- All terminating decimals (which have an end) or recurring decimals (which repeat forever) can be written as fractions.

Do you remember how to convert between fractions and decimals?

Changing a terminating decimal into a fraction

Follow these steps to change a terminating decimal into a fraction.

1 Write the decimal without the decimal point.
This will be the numerator of the fraction.
2 The denominator is a power of 10.
The number of zeros is the same as the number of decimal places in the original decimal.
3 Divide both the numerator and denominator by the biggest possible number (highest common factor, HCF). This gives the fraction in its **simplest form**.

Example

Write 0.12 as a fraction.

$0.12 = \frac{12}{100}$ Write 12 over 100 (two decimal places → two zeros).

$\quad\ = \frac{3}{25}$ Divide numerator and denominator by 4.

NOTE: If we have a mixed number (e.g. 1.5) then we leave the whole number as it is, and change only the decimal part into a fraction (so 1.5 becomes $1\frac{1}{2}$).

Changing a fraction into a decimal

To change a fraction into a decimal, divide the denominator into the numerator (that is, divide the numerator by the denominator).

Continue the division process until

- you have no remainder (the decimal is a terminating decimal), *or*
- you have 1 more decimal place than you are asked for (then round up or down), *or*
- you can clearly see the pattern of numbers that repeats in the decimal places (the decimal is a recurring decimal).

Example

Write $\frac{11}{20}$ as a decimal.

$$
\begin{array}{r}
0.55 \\
20\overline{)11.00} \\
\underline{10.0} \\
1.00 \\
\underline{1.00} \\
0
\end{array}
$$

$\frac{11}{20} = 0.55$

Revision

Recurring decimals

When we divide the denominator into the numerator of some fractions, the numbers in the decimal places make a **pattern of numbers** that **repeats** again and again.

NOTE: When something repeats again and again, we say that it **recurs**. So, any decimal that has a repeating pattern of numbers in the decimal places is called a **recurring decimal** (or sometimes a **repeating decimal**).

We use the symbol $\cdot$ to show recurring (or repeating) numbers in a decimal.

NOTE: Sometimes, in other books, you may see the symbol ⁻ instead.

If the **same number** recurs again and again, we write this number only once and put a single dot above this number.

For example:

$\frac{1}{3} = 0.333\,333\ldots = 0.\dot{3}$ $\frac{5}{6} = 0.833\,333\ldots = 0.8\dot{3}$

If **more than one** number recurs in the pattern, we write this recurring pattern of numbers only once, and put a dot above the first number in the pattern and a dot above the last number in the pattern.

For example:

$$\frac{4}{11} = 0.363636\ldots = 0.\dot{3}\dot{6}$$

$$\frac{8}{37} = 0.216216\ldots = 0.\dot{2}1\dot{6}$$

$$\frac{2}{7} = 0.285714285714\ldots = 0.\dot{2}8571\dot{4}$$

You can check all of the examples above by dividing the denominator into the numerator.

Any recurring decimal can be written as an exact fraction, $\frac{a}{b}$, where a is an integer, b is an integer, and $b \neq 0$. So we can say that:

Any recurring decimal is a rational number.

Check that you are confident with this revision work before going on with this unit.

Look at Exercises 7 and 8 in Unit 4 of Coursebook 1 and make sure you know how to answer all the questions in these exercises.

Changing a recurring decimal into a fraction

Follow these steps to change a recurring decimal into a fraction.

1 Let x = the recurring decimal.
(This step gives us equation 1.)

2 Count the number of digits in the repeating pattern of the recurring decimal.
Let this number be n.

3 Multiply both sides of equation 1 by 10^n (e.g. if there is only one recurring digit, we multiply by $10^1 = 10$; if there are two recurring digits, we multiply by $10^2 = 100$, and so on).
(This step gives us equation 2.)

4 Subtract equation 1 from equation 2 and solve the result to find x, which gives the fraction – remember to write it in its simplest form.

Examples

Find the fraction, in its simplest form, that is equal to each of these recurring decimals.

a) $0.\dot{4}$

Let $x = 0.444444\ldots$ (Equation 1)
Only 1 digit recurs, so we multiply by $10^1 = 10$.

$10x = 4.444444\ldots$ (Equation 2)

Subtract equation 1 from equation 2:

$$\begin{aligned} 10x &= 4.444444\ldots \\ -\quad x &= 0.444444\ldots \\ \hline 9x &= 4.0 \\ x &= \tfrac{4}{9} \end{aligned}$$

So $0.\dot{4} = \frac{4}{9}$

b) $0.\dot{5}\dot{7}$

Let $x = 0.575\,757...$ (Equation 1)
2 digits recur, so we multiply by $10^2 = 100$.
$100x = 57.575\,757...$ (Equation 2)
Subtract equation 1 from equation 2:

$$\begin{array}{r} 100x = 57.575\,757... \\ - \quad x = 0.575\,757... \\ \hline 99x = 57.0 \end{array}$$

$$x = \frac{57}{99}$$
$$= \frac{19}{33}$$

So $0.\dot{5}\dot{7} = \frac{19}{33}$

c) $0.1\dot{2}3\dot{4}$

Let $x = 0.123\,423\,423...$ (Equation 1)
3 digits recur, so we multiply by $10^3 = 1000$.
$1000x = 123.423\,423\,423...$ (Equation 2)
Subtract equation 1 from equation 2:

$$\begin{array}{r} 1000x = 123.423\,423\,423... \\ - \quad x = 0.123\,423\,423... \\ \hline 999x = 123.3 \end{array}$$

$$x = \frac{123.3}{999}$$
$$= \frac{1233}{9990} \qquad \text{(Because we can only have whole}$$
$$= \frac{137}{1110} \qquad\qquad \text{numbers in a fraction)}$$

So $0.1\dot{2}3\dot{4} = \frac{137}{1110}$

Exercise 1

1 Convert each of these recurring decimals to a fraction in its simplest form.

a) $0.\dot{8}$ b) $0.\dot{3}\dot{6}$ c) $0.4\dot{2}\dot{5}$ d) $0.\dot{2}857\dot{1}4$
e) $0.1\dot{6}$ f) $0.1\dot{8}\dot{2}$ g) $0.86\dot{1}$ h) $0.7\dot{7}2\dot{7}$

2 Convert each of these recurring decimals to a fraction in its simplest form.

a) $0.\dot{6}$ b) $0.\dot{2}\dot{1}$ c) $0.24\dot{5}$ d) $1.1\dot{4}\dot{7}$
e) $0.2\dot{3}$ f) $1.4\dot{5}\dot{7}$ g) $0.\dot{7}$ h) $0.\dot{4}\dot{5}$

3 Convert each of these recurring decimals to a fraction in its simplest form.

a) $0.2\dot{7}$ b) $0.8\dot{4}$ c) $0.4\dot{5}$ d) $0.5\dot{2}\dot{6}$
e) $2.4\dot{7}$ f) $1.3\dot{2}\dot{6}$ g) $0.8\dot{3}$ h) $4.3\dot{5}\dot{7}$

Irrational numbers

We know that any rational number can be written in the form of an exact fraction, $\frac{a}{b}$, where a is an integer, b is an integer and $b \neq 0$.

If this is true, then any number that **cannot** be written in the form of an exact fraction must be an irrational number.

Irrational numbers are **non-terminating** (meaning that they never end!), and **non-repeating** (so they do not have any recurring patterns of numbers) when we write them as **decimals**.

Common examples of irrational numbers are:

- Square roots of non-square numbers (that is, square roots of numbers that are not perfect squares)
- Cube roots of non-cube numbers (that is, cube roots of numbers that are not perfect cubes)

For example:

$$\sqrt{2} \qquad \sqrt{3} \qquad \sqrt[3]{7} \qquad \pi \qquad \sqrt{13} \qquad \text{are all irrational numbers.}$$

To make it easier in calculations, we often approximate commonly used irrational numbers to the required degree of accuracy by using an appropriate **rational** number.

Here are some **approximate** rational number values for some commonly used irrational numbers:

$$\sqrt{2} \approx 1.414 \qquad \sqrt{3} \approx 1.732 \qquad \pi \approx 3.142$$

Examples

a) State whether each of these numbers is rational or irrational. Explain your answers.

i) 0.425

Rational. 0.425 is a terminating decimal and so we can write it as a fraction. All terminating decimals can be written as a fraction, and so they are all rational numbers.

ii) $0.\dot{2}$

Rational. $0.\dot{2}$ is a recurring decimal. All recurring decimals can be written as a fraction, and so they are all rational numbers.

b) Work out whether each of these numbers is rational or irrational. Show your working.

i) 0.6

$$0.6 = \frac{6}{10} = \frac{3}{5}$$

So 0.6 is rational.

ii) π

$\pi = 3.141592\ldots$

π is non-terminating and non-recurring, with no exact value, so it is irrational.

iii) $\sqrt[3]{7}$

$\sqrt[3]{7} = 1.912931\ldots$

$\sqrt[3]{7}$ is non-terminating and non-recurring, with no exact value, so it is irrational.

iv) $\sqrt{36}$

$$\sqrt{36} = 6 = \frac{6}{1}$$

So $\sqrt{36}$ is rational.

v) $\sqrt{\dfrac{1}{25}}$

$$\sqrt{\frac{1}{25}} = \frac{\sqrt{1}}{\sqrt{25}} = \frac{1}{5}$$

So $\sqrt{\dfrac{1}{25}}$ is rational.

vi) $\sqrt{\dfrac{1}{72}}$

$$\sqrt{\frac{1}{72}} = \frac{\sqrt{1}}{\sqrt{72}} = \frac{1}{\sqrt{36 \times 2}} = \frac{1}{6\sqrt{2}}$$

$\sqrt{2}$ is irrational, so $\sqrt{\dfrac{1}{72}}$ cannot be written as an exact fraction, $\dfrac{a}{b}$, so it is irrational.

Exercise 2

1 These numbers are rational. To prove this, write each number as a fraction in its simplest form.

a) 0.5   b) 0.45   c) 0.3   d) 0.625   e) $2\frac{1}{4}$ f) -0.25

g) $\sqrt{49}$ h) $\dfrac{\sqrt{0.49}}{2}$ i) $\dfrac{\sqrt{64}}{\sqrt{16}}$ j) $-\dfrac{\sqrt{0.16}}{2}$ k) $\sqrt[3]{64}$ l) $\dfrac{\sqrt[3]{27}}{4}$

2 State whether each of these numbers is rational or irrational.

Write each of the rational numbers in the form $\frac{a}{b}$, where a and b are integers.

a) 0.75 b) $0.\dot{5}$ c) $\sqrt{5}$ d) $\sqrt{64}$ e) $\sqrt{6.4}$ f) $-\sqrt{0.64}$

g) $\frac{\pi}{2}$ h) $\frac{\sqrt{8}}{\sqrt{4}}$ i) $\sqrt[3]{8}$ j) 0.12 k) $\frac{1.8}{3}$ l) $3.\dot{1}$

3 State whether each of these numbers is rational or irrational.

Write each of the rational numbers in the form $\frac{a}{b}$, where a and b are integers.

a) -2π b) $\sqrt{2}$ c) 3.142 d) $\sqrt[3]{9}$ e) $\sqrt{\frac{1}{4}}$ f) $\frac{\sqrt{64}}{3}$

g) $\left(\frac{\sqrt{3}}{2}\right)^2$ h) $\sqrt{\frac{1}{2}}$ i) $\sqrt{1\frac{7}{9}}$ j) $\sqrt{6\frac{1}{4}}$ k) π^2 l) $\frac{\sqrt{5}}{2}$

m) $\left(\frac{\sqrt{5}}{2}\right)^2$ n) $\sqrt{2\frac{1}{4}}$ o) $\left(\frac{1}{3}\right)^2 + \frac{2}{3}$ p) $\sqrt[3]{-64}$

4 a) Write down one example of an irrational number that lies between 2 and 3.
   b) Write down one example of an irrational number that lies between 4 and 5.
   c) Write down one example of an irrational number that lies between 6 and 7.

5 m and n are two different irrational numbers.
Write down one example of values for m and n which make each of these statements true. (You will not be able to use the same values for each part.)

a) mn is rational.   b) mn is irrational.

c) $\frac{m}{n}$ is rational.   d) $\frac{m}{n}$ is irrational.

Rationalising denominators of fractions

Sometimes the denominator of a fraction is an irrational number in the form of a square root. In such cases we usually remove the square root from the denominator.

This process is called rationalising **the denominator** (because it changes the denominator from an irrational number into a rational number, but does not change the size of the fraction).

If we have a fraction in the form $\frac{a}{\sqrt{b}}$, we multiply both the numerator (top) and the denominator (bottom) of the fraction by the square root part of the denominator, $\sqrt{b}$. This does not change the value of the fraction because multiplying by $\frac{\sqrt{b}}{\sqrt{b}}$ is the same as multiplying by 1. After this, we can simplify the answer as much as possible.

Examples

Rationalise the denominator of each fraction and then simplify as much as possible.

a) $\dfrac{1}{\sqrt{3}}$

$$\dfrac{1}{\sqrt{3}} = \dfrac{1}{\sqrt{3}} \times \dfrac{\sqrt{3}}{\sqrt{3}}$$
$$= \dfrac{\sqrt{3}}{3}$$

b) $\dfrac{2\sqrt{3}}{\sqrt{6}}$

$$\dfrac{2\sqrt{3}}{\sqrt{6}} = \dfrac{2\sqrt{3}}{\sqrt{6}} \times \dfrac{\sqrt{6}}{\sqrt{6}}$$
$$= \dfrac{2\sqrt{3}\sqrt{6}}{6}$$
$$= \dfrac{2\sqrt{3}\sqrt{3}\sqrt{2}}{6}$$
$$= \dfrac{6\sqrt{2}}{6}$$
$$= \sqrt{2}$$

Exercise 3

1 Rationalise the denominator of each fraction and then simplify as much as possible.

a) $\dfrac{1}{\sqrt{7}}$ b) $\dfrac{1}{\sqrt{11}}$ c) $\dfrac{7}{\sqrt{7}}$ d) $\dfrac{10}{\sqrt{5}}$ e) $\dfrac{4}{\sqrt{8}}$ f) $\dfrac{21}{\sqrt{7}}$

g) $\dfrac{18}{\sqrt{6}}$ h) $\dfrac{7}{\sqrt{21}}$ i) $\dfrac{8}{\sqrt{24}}$ j) $\dfrac{14}{\sqrt{35}}$

2 Rationalise the denominator of each fraction and then simplify as much as possible.

a) $\dfrac{9}{\sqrt{18}}$ b) $\dfrac{6}{\sqrt{24}}$ c) $\dfrac{6}{\sqrt{8}}$ d) $\dfrac{\sqrt{5}}{\sqrt{20}}$ e) $\dfrac{\sqrt{12}}{\sqrt{3}}$ f) $\dfrac{\sqrt{15}}{\sqrt{5}}$

g) $\dfrac{2\sqrt{8}}{\sqrt{32}}$ h) $\dfrac{3\sqrt{5}}{\sqrt{15}}$ i) $\dfrac{\sqrt{50}}{\sqrt{75}}$ j) $\sqrt{\dfrac{9}{10}}$ k) $\dfrac{5}{2\sqrt{3}}$ l) $\dfrac{\sqrt{3}\sqrt{5}}{\sqrt{30}}$

NOTE: If you study maths beyond school level, you will discover another group of numbers, called imaginary numbers. These are numbers like $\sqrt{-1}$ which, for now, we say do not exist!

Unit 7 Indices

Key vocabulary

base	integral index	radical sign
fractional index	power	rational index
index	radical expression	reciprocal
indices	radical form	

A Indices

Revision

In Unit 5 of Coursebook 1 we learned about power numbers. We also learned that the correct name for a power number is index (plural: indices).

In Unit 9 of Coursebook 1 we learned to work with indices that have a **variable** as the base number.

Remember:

- If we have any number, a, then $a \times a \times a \times a \times a$ can be written in index form as a^5.

  a^5 is read as 'a to the power 5'.

  a is the **base** of the power number.

  5 is the **power** or **index**.

- There are two special power numbers.

  - In general, for any number a, $a^0 = 1$. That is, any number to the power of zero is equal to 1.

    So $2^0 = 1$, $3^0 = 1$, $4^0 = 1$, $5^0 = 1$ and so on.

    This is sometimes called the **zero index rule**.

  - In general, for any number a, $a^1 = a$. That is, any number to the power of 1 is equal to that same number.

    So $2^1 = 2$, $3^1 = 3$, $4^1 = 4$, $5^1 = 5$ and so on.

- When numbers (or variables) written in index form with the **same** base are **multiplied**, the indices are **added**:

$$a^m \times a^n = a^{m+n}$$

This is called the **first law of indices**.

- When numbers (or variables) written in index form with the **same base** are **divided**, the indices are **subtracted**:

$$a^m \div a^n = \frac{a^m}{a^n} = a^{m-n}, \text{ where } a \neq 0$$

This is called the **second law of indices**.

Examples

a) Use the first law of indices to simplify each of these calculations:

i) $6^5 \times 6^4$

$$6^5 \times 6^4 = 6^{5+4} = 6^9$$

ii) $p^2 \times p^6$

$$p^2 \times p^6 = p^{2+6} = p^8$$

b) Use the second law of indices to simplify each of these calculations.

i) $\dfrac{5^7}{5^3}$

$$\frac{5^7}{5^3} = 5^7 \div 5^3$$
$$= 5^{7-3} = 5^4$$

ii) $x^8 \div x^5$

$$x^8 \div x^5 = x^{8-5} = x^3$$

Remember that the index laws can only be used if the base numbers are the **same**.
When we multiply or divide indices with different base numbers, each base number must be calculated **separately**.

Examples

a) Use the first law of indices to simplify each of these calculations.

i) $3^4 \times 2^3 \times 3^7 \times 2^5$

$$3^4 \times 2^3 \times 3^7 \times 2^5 = 3^4 \times 3^7 \times 2^3 \times 2^5$$
$$= 3^{4+7} \times 2^{3+5}$$
$$= 3^{11} \times 2^8$$

ii) $2a^3b^2 \times 3ab^4$

$$2a^3b^2 \times 3ab^4 = 2 \times a^3 \times b^2 \times 3 \times a \times b^4$$
$$= (2 \times 3) \times (a^3 \times a) \times (b^2 \times b^4)$$
$$= 6 \times a^{3+1} \times b^{2+4}$$
$$= 6 \times a^4 \times b^6$$
$$= 6a^4b^6$$

b) Use the first and second laws of indices to simplify each of these calculations.

i) $\dfrac{2^5 \times 5^3 \times 2^3 \times 5^4}{5^2 \times 2^2}$

$$\dfrac{2^5 \times 5^3 \times 2^3 \times 5^4}{5^2 \times 2^2} = \dfrac{2^5 \times 2^3}{2^2} \times \dfrac{5^3 \times 5^4}{5^2}$$
$$= 2^{5+3-2} \times 5^{3+4-2}$$
$$= 2^6 \times 5^5$$

ii) $\dfrac{8x^5y^6}{2x^4y^3}$

$$\dfrac{8x^5y^6}{2x^4y^3} = \dfrac{8}{2} \times \dfrac{x^5}{x^4} \times \dfrac{y^6}{y^3}$$
$$= 4 \times x^{5-4} \times y^{6-3}$$
$$= 4 \times x^1 \times y^3$$
$$= 4xy^3$$

Exercise 1

1 Use the first law of indices to simplify each of these calculations.

a) $2^2 \times 2^6$ 　　　b) $7^3 \times 7^7$ 　　　c) $9^5 \times 9^6$
d) $b^5 \times b^5$ 　　　e) $s^3 \times s^{14}$ 　　　f) $x^7 \times x^5$

2 Use the second law of indices to simplify each of these calculations.

a) $3^{13} \div 3^8$ 　　　b) $\dfrac{4^9}{4^4}$ 　　　c) $\dfrac{5^{19}}{5^{11}}$

d) $q^{11} \div q^8$ 　　　e) $\dfrac{t^{12}}{t^9}$ 　　　f) $\dfrac{g^8}{g^7}$

3 Simplify each of these calculations using the first law of indices.

a) $2x^5 \times 10x^9$ 　　　b) $8m^4 \times 3m^9$ 　　　c) $4g^5 \times 3g^7$
d) $a^2b^3 \times b^4$ 　　　e) $m^2n^2 \times m^4n^5$ 　　　f) $b^5k^4 \times b^2k^3$
g) $8a^2b^2 \times 2a^5b^2$ 　　　h) $5p^2q \times 3pq^3$ 　　　i) $3m^3n^4 \times 4m^4n^3$

4 Simplify each of these calculations using the laws of indices.

a) $\dfrac{32a^5}{24a^2}$

b) $\dfrac{25p^4}{15p}$

c) $\dfrac{4m^4n^7}{8m^3n^4}$

d) $\dfrac{14s^{14}t^8}{35s^9t^5}$

e) $\dfrac{2^6 \times 3^7 \times 2^3 \times 3^3}{2^2 \times 3^5}$

f) $\dfrac{4^5 \times 9^6 \times 4^6 \times 9^7}{4^2 \times 9^4 \times 4^4 \times 9^5}$

g) $\dfrac{2x^4 \times 3n^3 \times 3x^2 \times 5n^2}{n^4 \times x^5}$

h) $\dfrac{2t^3 \times a^7 \times t^9 \times 10a^4}{t^4 \times 6a^3 \times 5t^6 \times a^5}$

i) $\dfrac{c^8 \times 8p^{12} \times 3c^7 \times p^5}{6c^3 \times p^6 \times c^5 \times p^7}$

B The third law of indices

Look at this calculation: $(5^2)^3$.

We know that the index 3 outside the brackets means that everything inside the brackets is multiplied by itself 3 times.

So $(5^2)^3 = 5^2 \times 5^2 \times 5^2$
$= 5^{2+2+2}$ (Using the first law of indices)
$= 5^6$

We know that $2 + 2 + 2 = 2 \times 3$, so we can see that the answer is found by **multiplying** the indices.

In the same way, $(y^5)^4 = y^5 \times y^5 \times y^5 \times y^5$
$= y^{5+5+5+5}$
$= y^{5 \times 4} = y^{20}$

> When numbers (or variables) written in index form with the **same base** are raised to another power, the indices are **multiplied**:
>
> $$(a^m)^n = a^{(m \times n)} = a^{mn}$$
>
> This is called the **third law of indices**.

Examples

Use the third law of indices to simplify each of these calculations.

a) $(4^4)^4$

$(4^4)^4 = 4^{4 \times 4}$
$= 4^{16}$

b) $(g^5)^6$

$(g^5)^6 = g^{5 \times 6}$
$= g^{30}$

In Coursebook 1 we learned about the **order of operations** when working with integers and fractions. When working with indices, we follow the same order of operations. This means that we always **simplify brackets first** before multiplying or dividing indices.

Examples

Use the laws of indices to write each of these in its simplest index form.

a) $(b^3)^2 \times b^7$

$$\begin{aligned}(b^3)^2 \times b^7 &= (b^{3 \times 2}) \times b^7 \\ &= b^6 \times b^7 \\ &= b^{6+7} \\ &= b^{13}\end{aligned}$$

b) $d^{10} \div (d^2)^5$

$$\begin{aligned}d^{10} \div (d^2)^5 &= d^{10} \div (d^{2 \times 5}) \\ &= d^{10} \div d^{10} \\ &= d^{10-10} \\ &= d^0 = 1\end{aligned}$$

c) $(p^4)^2 \times (p^2)^3$

$$\begin{aligned}(p^4)^2 \times (p^2)^3 &= (p^{4 \times 2}) \times (p^{2 \times 3}) \\ &= p^8 \times p^6 \\ &= p^{8+6} \\ &= p^{14}\end{aligned}$$

d) $\dfrac{(z^3)^4}{(z^2)^5}$

$$\begin{aligned}\frac{(z^3)^4}{(z^2)^5} &= \frac{z^{3 \times 4}}{z^{2 \times 5}} \\ &= \frac{z^{12}}{z^{10}} \\ &= z^{12-10} \\ &= z^2\end{aligned}$$

Exercise 2

1 Use the third law of indices to write each of these in the simplest index form.

a) $(3^2)^4$ b) $(8^4)^5$ c) $(a^6)^3$

d) $(m^3)^2$ e) $(p^5)^5$ f) $(s^4)^4$

g) $(b^8)^4$ h) $(n^6)^3$

2 Use the laws of indices to write each of these in its simplest index form.

a) $(a^2)^2 \times a^4$

b) $(m^4)^3 \times m^2$

c) $s^2 \times (s^3)^2$

d) $t^5 \times (t^2)^4$

e) $b \times (b^3)^3$

3 Use the laws of indices to write each of these in its simplest index form.

a) $(a^6)^2 \div a^3$

b) $(b^8)^6 \div b^{10}$

c) $(y^6)^3 \div y$

d) $(g^2)^3 \div g^3$

e) $(s^2)^2 \div s^4$

4 Use the laws of indices to write each of these in its simplest index form.

a) $(a^5)^2 \times (a^4)^3$

b) $(m^4)^4 \times (m^2)^3$

c) $(p^2)^3 \times (p^3)^2$

d) $(c^5)^4 \div (c^4)^3$

e) $(a^3)^7 \div (a^5)^3$

f) $(y^4)^3 \div (y^6)^2$

C The fourth law of indices

Look at this calculation: $(4 \times 3)^3$.

We know that the index 3 means that (4×3) is multiplied by itself three times.

So $(4 \times 3)^3 = (4 \times 3) \times (4 \times 3) \times (4 \times 3)$

$$= 4 \times 4 \times 4 \times 3 \times 3 \times 3$$
$$= 4^3 \times 3^3$$

So we can say that $(4 \times 3)^3 = 4^3 \times 3^3$.

In the same way, $(st)^4 = (s \times t)^4$

$$= (s \times t) \times (s \times t) \times (s \times t) \times (s \times t)$$
$$= s \times s \times s \times s \times t \times t \times t \times t$$
$$= s^4 \times t^4$$

So we can say that $(st)^4 = s^4t^4$.

> When numbers (or variables) are multiplied together and then raised to a power, the answer is the same as raising each number (or variable) to this power before multiplying:
>
> $$(a \times b)^n = a^n \times b^n \quad or \quad (ab)^n = a^nb^n$$
>
> This is called the **fourth law of indices**.

Examples   Use the laws of indices to simplify each of these.

a) $(m^4 \times n)^3$

$$(m^4 \times n)^3 = (m^4)^3 \times n^3$$
$$= m^{4\times3} \times n^3$$
$$= m^{12} \times n^3$$
$$= m^{12}n^3$$

b) $(3xy^3)^2$

$$(3xy^3)^2 = (3 \times x \times y^3)^2$$
$$= 3^2 \times x^2 \times (y^3)^2$$
$$= 3^2 \times x^2 \times y^{3 \times 2}$$
$$= 9x^2y^6$$

c) $(7p^4rs^5)^0$

$(7p^4rs^5)^0 = 1$ (Because anything to the power 0 is equal to 1)
Check:
$$(7p^4rs^5)^0 = (7 \times p^4 \times r \times s^5)^0$$
$$= 7^0 \times p^{4 \times 0} \times r^0 \times s^{5 \times 0}$$
$$= 1 \times 1 \times 1 \times 1$$
$$= 1$$

Exercise 3

1 Use the fourth law of indices to simplify each of these.
   a) $(cd)^6$
   b) $(4b)^3$
   c) $(7d)^3$
   d) $(5rs)^4$
   e) $(4cd)^2$
   f) $(17fg)^0$

2 Use the laws of indices to simplify each of these.
   a) $(c^3d)^2$
   b) $(c^2d^2)^4$
   c) $(m^3n^2)^5$
   d) $(r^4s^3)^4$
   e) $(3e^2f^3)^3$
   f) $(2g^4h^2)^4$
   g) $(3y^2x^3)^5$
   h) $(23p^7q^3)^0$

D The fifth law of indices

Look at this calculation: $\left(\frac{3}{4}\right)^3$.

We know that the index 3 means that $\frac{3}{4}$ is multiplied by itself three times.

So $\left(\dfrac{3}{4}\right)^3 = \dfrac{3}{4} \times \dfrac{3}{4} \times \dfrac{3}{4}$

$$= \frac{3 \times 3 \times 3}{4 \times 4 \times 4}$$

$$= \frac{3^3}{4^3}$$

So we can say that $\left(\dfrac{3}{4}\right)^3 = \dfrac{3^3}{4^3}$.

In the same way, $\left(\dfrac{c}{d}\right)^4 = \dfrac{c}{d} \times \dfrac{c}{d} \times \dfrac{c}{d} \times \dfrac{c}{d}$

$$= \dfrac{c \times c \times c \times c}{d \times d \times d \times d}$$

$$= \dfrac{c^4}{d^4}$$

So we can say that $\left(\dfrac{c}{d}\right)^4 = \dfrac{c^4}{d^4}$.

When numbers (or variables) are divided and then raised to a power, the answer is the same as raising each number (or variable) to this power before dividing:

$$(a \div b)^n = a^n \div b^n \quad or \quad \left(\dfrac{a}{b}\right)^n = \dfrac{a^n}{b^n}$$

This is called the **fifth law of indices**.

Examples

Use the laws of indices to simplify each of these.

a) $\left(\dfrac{p}{r}\right)^7$

$\left(\dfrac{p}{r}\right)^7 = \dfrac{p^7}{r^7}$

b) $\left(\dfrac{4g^2h}{k^4}\right)^3$

$\left(\dfrac{4g^2h}{k^4}\right)^3 = \dfrac{(4g^2h)^3}{(k^4)^3}$

$$= \dfrac{4^3 g^{2\times3} h^3}{k^{4\times3}}$$

$$= \dfrac{64g^6h^3}{k^{12}}$$

 Exercise 4

1 Use the laws of indices to simplify each of these.

a) $\left(\dfrac{5}{6}\right)^2$

b) $\left(\dfrac{4}{9}\right)^3$

c) $\left(\dfrac{c}{d}\right)^5$

d) $\left(\dfrac{s}{t}\right)^6$

e) $\left(\dfrac{8a}{b}\right)^3$

2 Use the laws of indices to simplify each of these.

a) $\left(\dfrac{c^2}{d}\right)^2$

b) $\left(\dfrac{e^3}{f}\right)^2$

c) $\left(\dfrac{k}{m^4}\right)^5$

d) $\left(\dfrac{x^3}{y^2}\right)^5$

e) $\left(\dfrac{g^8}{b^6}\right)^3$

f) $\left(\dfrac{2a^2b}{c^2}\right)^3$

g) $\left(\dfrac{5pr^3}{q^4}\right)^2$

h) $\left(\dfrac{3b^3k^4}{2g^2}\right)^5$

i) $\left(\dfrac{11c^3d^7}{13e^8}\right)^2$

j) $\left(\dfrac{7m^5n}{p^2r^3}\right)^3$

E Negative indices

In Unit 5 of Coursebook 1 we learned that indices can also be negative numbers.

Look at $\dfrac{1}{7^3}$. We know that any number to the power of 0 is equal to 1.

So we can write $1 = 7^0$.

$$\dfrac{1}{7^3} = \dfrac{7^0}{7^3}$$

$$= 7^0 \div 7^3 = 7^{0-3} = 7^{-3}$$

So we can say that $\dfrac{1}{7^3} = 7^{-3}$.

We can also look at this another way.

$$\dfrac{a^2}{a^4} = \dfrac{a \times a}{a \times a \times a \times a} \quad \text{and} \quad \dfrac{a^2}{a^4} = a^2 \div a^4$$

$$= \dfrac{1}{a^2} \qquad\qquad\qquad = a^{2-4} \quad \text{(Using the second law}$$

$$= a^{-2} \quad \text{of indices)}$$

So we can say that $a^{-2} = \dfrac{1}{a^2}$.

In general we can say that $a^{-n} = \dfrac{1}{a^n}$, where $a \neq 0$. So a^{-n} is the reciprocal of a^n, $\dfrac{1}{a^n}$.

All the five laws of indices that we have used with **positive indices** are also true for **negative indices**.

Examples

a) Use the laws of indices to simplify each of these. Write the answers using positive indices.

i) d^{-6}

$$d^{-6} = \frac{1}{d^6}$$

ii) $3c^{-4}$

$$3c^{-4} = 3 \times c^{-4} \quad \text{(3 is not raised to the power of } -4.\text{)}$$
$$= 3 \times \frac{1}{c^4}$$
$$= \frac{3}{c^4}$$

iii) $s^3 \times s^{-5}$

$$s^3 \times s^{-5} = s^{3+(-5)}$$
$$= s^{3-5}$$
$$= s^{-2}$$
$$= \frac{1}{s^2}$$

iv) $b^{-3} \div b^8$

$$b^{-3} \div b^8 = b^{-3-(8)}$$
$$= b^{-11}$$
$$= \frac{1}{b^{11}}$$

v) $4(y^4)^{-3}$

$$4(y^4)^{-3} = 4 \times y^{4 \times (-3)} \quad \text{(4 is not raised to the power of } -3.\text{)}$$
$$= 4y^{-12}$$
$$= \frac{4}{y^{12}}$$

vi) $(4y^4)^{-3}$

$$(4y^4)^{-3} = 4^{-3}y^{4 \times (-3)}$$
$$= \frac{1}{4^3} \times \frac{1}{y^{12}}$$
$$= \frac{1}{64y^{12}}$$

b) Use the laws of indices to simplify each of these. Write the answers using positive indices.

i) $t^{-2} \times t^{-5} \div t^2$

$$t^{-2} \times t^{-5} \div t^2 = t^{-2+(-5)-(2)}$$
$$= t^{-2-5-2}$$
$$= t^{-9}$$
$$= \frac{1}{t^9}$$

ii) $\dfrac{(a^2)^{-5} \times (ab)^4}{(a^{-1}b)^2}$

$$\frac{(a^2)^{-5} \times (ab)^4}{(a^{-1}b)^2} = \frac{a^{2\times(-5)} \times a^4 \times b^4}{a^{(-1)\times 2} \times b^2}$$
$$= \frac{a^{-10+4} \times b^4}{a^{-2} \times b^2}$$
$$= a^{-6-(-2)} \times b^{4-2}$$
$$= a^{-4} \times b^2$$
$$= \frac{b^2}{a^4}$$

c) Use the laws of indices to simplify each of these. Write the answers using negative indices.

i) $\dfrac{(a^2b)^3}{a^7b^5}$

$$\frac{(a^2b)^3}{a^7b^5} = \frac{(a^2)^3 b^3}{a^7b^5}$$
$$= \frac{a^6 b^3}{a^7 b^5}$$
$$= a^{6-7}b^{3-5}$$
$$= a^{-1}b^{-2}$$

ii) $\dfrac{4a^{-3}b^2}{(2ab^2)^3}$

$$\frac{4a^{-3}b^2}{(2ab^2)^3} = \frac{4a^{-3}b^2}{8a^3b^6}$$
$$= \frac{1}{2} \times a^{-3-3}b^{2-6}$$
$$= \frac{1}{2}a^{-6}b^{-4}$$

NOTE: We normally write all numbers with positive indices (so in Example **c)** part **ii)** we have $\frac{1}{2}$ not 2^{-1}), but variables can be written with positive or negative indices as required.

d) Evaluate each of these numbers.

   i) 3^{-3}

$$3^{-3} = \frac{1}{3^3}$$

$$= \frac{1}{27}$$

   ii) $2^3 \times 3^{-2}$

$$2^3 \times 3^{-2} = 2^3 \times \frac{1}{3^2}$$

$$= \frac{8}{9}$$

   iii) $\left(\frac{2}{3}\right)^{-2} \times \left(\frac{7}{11}\right)^0$

$$\left(\frac{2}{3}\right)^{-2} \times \left(\frac{7}{11}\right)^0 = \frac{1}{\left(\frac{2}{3}\right)^2} \times 1$$

$$= \frac{1}{\frac{4}{9}}$$

$$= \frac{9}{4}$$

$$= 2\frac{1}{4}$$

   iv) $(3^{-2})^3 \times (9^{-3})^{-2}$

$$(3^{-2})^3 \times (9^{-3})^{-2} = 3^{-6} \times 9^6$$
$$= 3^{-6} \times (3^2)^6 \quad \text{(Change to the same base.)}$$
$$= 3^{-6} \times 3^{12}$$
$$= 3^6$$
$$= 729$$

NOTE: Example **d)** part **iii)** shows us another quick rule:

$$\left(\frac{a}{b}\right)^{-n} = \left(\frac{b}{a}\right)^n, \text{ where } a \neq 0 \text{ and } b \neq 0.$$

Exercise 5

1 Write each of these with positive indices.

a) b^{-3} b) n^{-6} c) $4b^{-4}$
d) $5c^{-5}$ e) $3^{-3}p^{-2}$ f) $5^{-4}d^{-7}$

2 Use the laws of indices to simplify each of these. Write the answers using positive indices.

a) $3^3 \times 3^{-7}$ b) $10^{-4} \times 10^{-2}$ c) $r^{-5} \times r$
d) $3b^{-1} \times 4b^{-4}$ e) $q^{-2} \times q^2$ f) $c^2d^3 \times c^{-3}d^{-2}$

3 Use the laws of indices to simplify each of these. Write the answers using positive indices.

a) $3^3 \div 3^{-7}$ b) $5^{-1} \div 5^{-3}$ c) $m^5 \div m^{-1}$
d) $8g^3 \div 4g^4$ e) $\dfrac{5}{t} \div t^{-2}$ f) $\dfrac{9c^{-4}}{3c^6}$

4 Use the laws of indices to simplify each of these. Write the answers using positive indices.

a) $(5^{-4})^3$ b) $(a^2)^{-3}$ c) $2a^{-5} \div 7b^{-5}$
d) $\dfrac{1}{a^3b^{-4}}$ e) $\dfrac{2}{(a^3b^{-2})^2}$ f) $\dfrac{(ab^3)^{-4}}{a^{-3}b^{-7}}$
g) $\dfrac{(a^4)^3(a^{-1}b)^{10}}{a^2b^7}$ h) $(2a^2)^3 \times (4ab^{-1})^{-2}$ i) $\dfrac{(a^2b^{-1})^2}{a^3b^{-5}}$

5 Use the laws of indices to simplify each of these. Write the answers using negative indices.

a) $4^4 \times 4^{-2}$ b) $7^0 \times 7^{-4}$ c) $(2^{-2})^{-4}$
d) $c^2 \times d^3$ e) $(a^3b)^2$ f) $d^2 \div d^5$
g) $abc \div a^5b^4c^2$ h) $\dfrac{1}{a^5b^{-2}}$ i) $\dfrac{a^3b^2}{(ab)^{-2}}$
j) $\dfrac{a^3b^{-4}}{ab^2}$ k) $\left(\dfrac{a}{b^{-4}}\right)^{-2}$ l) $\left(\dfrac{-b^2}{a}\right)^2 \times a^{-4}$

6 Evaluate each of these. (Use the laws of indices to help you.)

a) 5^{-2} b) $(53)^0$ c) 3^{-3}
d) $\left(\dfrac{-1}{4}\right)^2$ e) $\left(\dfrac{2}{5}\right)^{-3}$ f) $\left(\dfrac{16}{9}\right)^{-2}$
g) $3^2 \times 4^{-3}$ h) $\left(\dfrac{3}{4}\right)^{-1} \times \left(\dfrac{3}{8}\right)^2$ i) $\dfrac{2^3 \times 6^{-5}}{3^{-3} \times 4^{-4}}$
j) $2^5 \times 2^3 \div 4^4$ k) $\left(\dfrac{3}{4}\right)^{-2} \div \left(\dfrac{4}{9}\right)^3 \times \left(\dfrac{27}{16}\right)^{-1}$ l) $\dfrac{5^4}{3^7} \times \left(\dfrac{9}{15}\right)^3 \div \dfrac{27}{25}$

Fractional indices

So far, we have learned about indices that are integers – these are called integral indices.

Some indices are in the form of fractions – these are called fractional indices. Also, because we know that any rational number can be written as a fraction, fractional indices are sometimes called rational indices.

The first law of indices is true for factional indices.

So $a^{\frac{1}{2}} \times a^{\frac{1}{2}} = a^{\frac{1}{2}+\frac{1}{2}}$
$\qquad\qquad = a^1$
$\qquad\qquad = a$

We also know that anything multiplied by itself makes a perfect square.

So $a^{\frac{1}{2}} \times a^{\frac{1}{2}} = \left(a^{\frac{1}{2}}\right)^2$

This means that we can say $\left(a^{\frac{1}{2}}\right)^2 = a$.

If we take the square root of both sides, we get $a^{\frac{1}{2}} = \sqrt{a}$.

Now let's look at $x^{\frac{1}{3}}$ in the same way.
$x^{\frac{1}{3}} \times x^{\frac{1}{3}} \times x^{\frac{1}{3}} = x^{\frac{1}{3}+\frac{1}{3}+\frac{1}{3}}$
$\qquad\qquad\qquad = x^1$
$\qquad\qquad\qquad = x$

We know that anything multiplied by itself three times makes a perfect cube.

So $x^{\frac{1}{3}} \times x^{\frac{1}{3}} \times x^{\frac{1}{3}} = \left(x^{\frac{1}{3}}\right)^3$

So we can say that $\left(x^{\frac{1}{3}}\right)^3 = x$.

If we take the cube root of both sides, we get $x^{\frac{1}{3}} = \sqrt[3]{x}$.

> In general we can say that $a^{\frac{1}{n}} = \sqrt[n]{a}$, where $a > 0$ and n is a positive integer.

NOTE: Even though we have studied only square roots and cube roots in detail (Unit 5), it is also possible to find **fourth roots** ($\sqrt[4]{\ }$), **fifth roots** ($\sqrt[5]{\ }$), **sixth roots** ($\sqrt[6]{\ }$), and so on …

Now let's look at a **fractional index** where the numerator is an integer > 1. For example, $a^{\frac{2}{3}}$.

Using the third law of indices, $\left(a^{\frac{2}{3}}\right)^3 = a^{\frac{2}{3} \times 3}$

$$= a^2$$

So $a^{\frac{2}{3}} = \sqrt[3]{a^2}$ (Take the cube root of both sides.)

We also know that $a^{\frac{2}{3}} = a^{\frac{1}{3} \times 2} = \left(a^{\frac{1}{3}}\right)^2$ and that $a^{\frac{1}{3}} = \sqrt[3]{a}$

So $a^{\frac{2}{3}} = (\sqrt[3]{a})^2$ and $\sqrt[3]{a^2} = (\sqrt[3]{a})^2$.

> In general we can say that $a^{\frac{m}{n}} = (\sqrt[n]{a})^m = \sqrt[n]{a^m}$, where $a > 0$ and m and n are integers.

NOTES:

- We know that $\sqrt{} = \sqrt[2]{}$ is called the **square root sign** and $\sqrt[3]{}$ is the **cube root sign**.
  In general, $\sqrt[n]{}$ is called the radical sign, when n is a positive integer.
- Any expression that includes the radical sign $\sqrt[n]{}$ is written in radical form and is called a radical expression.
- All five laws of indices are true for **integral indices** and for **rational indices**.

Examples

a) Simplify each of these.

  i) $64^{\frac{2}{3}}$

$$64^{\frac{2}{3}} = (4^3)^{\frac{2}{3}}$$
$$= 4^{3 \times \frac{2}{3}}$$
$$= 4^2$$
$$= 16$$

  ii) $(a^8)^{\frac{1}{2}}$

$$(a^8)^{\frac{1}{2}} = a^{8 \times \frac{1}{2}}$$
$$= a^4$$

b) Write each of these in radical form and then evaluate.

  i) $4^{\frac{1}{2}}$

$$4^{\frac{1}{2}} = \sqrt{4}$$
$$= 2$$

  ii) $27^{-\frac{1}{3}}$

$$27^{-\frac{1}{3}} = \frac{1}{\sqrt[3]{27}}$$
$$= \frac{1}{3}$$

iii) $8^{\frac{2}{3}}$

$$8^{\frac{2}{3}} = \sqrt[3]{8^2}$$
$$= \sqrt[3]{64}$$
$$= 4$$

iv) $49^{1.5}$

$$49^{1.5} = 49^{\frac{3}{2}}$$
$$= (\sqrt{49})^3$$
$$= 7^3$$
$$= 343$$

c) Write each of these radical expressions using rational indices.

i) $\sqrt{d^7}$

$$\sqrt{d^7} = (d^7)^{\frac{1}{2}}$$
$$= d^{\frac{7}{2}}$$

ii) $\sqrt[5]{b^2}$

$$\sqrt[5]{b^2} = (b^2)^{\frac{1}{5}}$$
$$= b^{\frac{2}{5}}$$

iii) $\dfrac{1}{\sqrt[s]{y^m}}$

$$\frac{1}{\sqrt[s]{y^m}} = \frac{1}{(y^m)^{\frac{1}{s}}}$$
$$= \frac{1}{y^{\frac{m}{s}}}$$
$$= y^{-\frac{m}{s}}$$

iv) $\dfrac{1}{\sqrt[y]{y^c}}$

$$\frac{1}{\sqrt[y]{y^c}} = \frac{1}{(y^c)^{\frac{1}{y}}}$$
$$= \frac{1}{y^{\frac{c}{y}}}$$
$$= y^{-\frac{c}{y}}$$

d) Use the rules of indices to simplify each of these. Write the answers with rational indices.

i) $p^{\frac{2}{3}} \times p^{\frac{1}{4}}$

$$p^{\frac{2}{3}} \times p^{\frac{1}{4}} = p^{\frac{2}{3}+\frac{1}{4}}$$
$$= p^{\frac{11}{12}}$$

ii) $7p^{\frac{3}{5}} \div p^{\frac{1}{2}}$

$$7p^{\frac{3}{5}} \div p^{\frac{1}{2}} = 7p^{\frac{3}{5}-\frac{1}{2}}$$
$$= 7p^{\frac{1}{10}}$$

iii) $\left(p^{\frac{5}{6}}\right)^{\frac{8}{15}}$

$$\left(p^{\frac{5}{6}}\right)^{\frac{8}{15}} = p^{\frac{5}{6} \times \frac{8}{15}}$$
$$= p^{\frac{4}{9}}$$

iv) $\left(p^{\frac{1}{3}}r^{\frac{2}{5}}\right)^{\frac{3}{4}}$

$$\left(p^{\frac{1}{3}}r^{\frac{2}{5}}\right)^{\frac{3}{4}} = p^{\frac{1}{3} \times \frac{3}{4}}r^{\frac{2}{5} \times \frac{3}{4}}$$
$$= p^{\frac{1}{4}}r^{\frac{3}{10}}$$

e) Use the rules of indices to simplify each of these. Write the answers in radical form.

NOTE: We learned in Unit 5 to simplify calculations with square roots – now we can extend this to any radical expression.

i) $\sqrt{3} \times \sqrt{5}$

$$\sqrt{3} \times \sqrt{5} = 3^{\frac{1}{2}} \times 5^{\frac{1}{2}}$$
$$= (3 \times 5)^{\frac{1}{2}} = \sqrt{15}$$

ii) $\sqrt[3]{2} \times \sqrt[3]{11}$

$$\sqrt[3]{2} \times \sqrt[3]{11} = 2^{\frac{1}{3}} \times 11^{\frac{1}{3}}$$
$$= (2 \times 11)^{\frac{1}{3}} = \sqrt[3]{22}$$

iii) $\sqrt[4]{4} \times \sqrt[6]{16}$

$$\sqrt[4]{4} \times \sqrt[6]{16} = 4^{\frac{1}{4}} \times 16^{\frac{1}{6}}$$
$$= (2^2)^{\frac{1}{4}} \times (2^4)^{\frac{1}{6}}$$
$$= 2^{\frac{1}{2}} \times 2^{\frac{2}{3}}$$
$$= 2^{\frac{7}{6}} = \sqrt[6]{2^7}$$

iv) $\sqrt{50} \times \sqrt[7]{64}$

$$\sqrt{50} \times \sqrt[7]{64} = 50^{\frac{1}{2}} \times 64^{\frac{1}{7}}$$
$$= (25 \times 2)^{\frac{1}{2}} \times (2^6)^{\frac{1}{7}}$$
$$= 25^{\frac{1}{2}} \times 2^{\frac{1}{2}} \times 2^{\frac{6}{7}}$$
$$= 5 \times 2^{\frac{19}{14}} = 5 \times \sqrt[14]{2^{19}}$$

f) Use the rules of indices to simplify each of these. Write the answers in radical form.

i) $g^{\frac{1}{2}} \times g^{\frac{2}{3}} \div g^{\frac{3}{4}}$

$$g^{\frac{1}{2}} \times g^{\frac{2}{3}} \div g^{\frac{3}{4}} = g^{\frac{1}{2}+\frac{2}{3}-\frac{3}{4}}$$
$$= g^{\frac{5}{12}} = \sqrt[12]{g^5}$$

ii) $\left(g^{\frac{1}{2}}b^{\frac{2}{3}}\right)^{\frac{3}{4}} \div \left(g^{\frac{2}{5}}b^{\frac{1}{3}}\right)^{\frac{5}{6}}$

$$\left(g^{\frac{1}{2}}b^{\frac{2}{3}}\right)^{\frac{3}{4}} \div \left(g^{\frac{2}{5}}b^{\frac{1}{3}}\right)^{\frac{5}{6}} = \left(g^{\frac{1}{2}\times\frac{3}{4}}b^{\frac{2}{3}\times\frac{3}{4}}\right) \div \left(g^{\frac{2}{5}\times\frac{5}{6}}b^{\frac{1}{3}\times\frac{5}{6}}\right)$$
$$= \left(g^{\frac{3}{8}}b^{\frac{1}{2}}\right) \div \left(g^{\frac{1}{3}}b^{\frac{5}{18}}\right)$$
$$= g^{\frac{3}{8}-\frac{1}{3}}b^{\frac{1}{2}-\frac{5}{18}}$$
$$= g^{\frac{1}{24}}b^{\frac{2}{9}} = \sqrt[24]{g} \times \sqrt[9]{b^2}$$

Exercise 6

1 Write each of these radical expressions using rational indices.

a) $\sqrt[3]{c}$

b) $\sqrt[4]{p}$

c) $\sqrt[3]{a^2}$

d) $\sqrt{b^3}$

e) $\sqrt[5]{k^6}$

f) $\sqrt{a^{16}}$

g) $\sqrt[3]{27e^9}$

h) $\sqrt[4]{y^{32}}$

i) $\sqrt[4]{81a^{20}}$

j) $\sqrt[3]{125m^9n^{15}}$

k) $\sqrt[3]{\dfrac{x^5y^6}{z^4}}$

l) $\sqrt[r]{\dfrac{a^v}{b^w}}$

2 Use the rules of indices to simplify each of these. Write the answers with rational indices.

a) $\left(x^6\right)^{\frac{1}{3}}$

b) $\left(r^{\frac{1}{2}}\right)^8$

c) $\left(p^{\frac{1}{3}}\right)^2$

d) $\left(2u^{\frac{1}{2}}\right)^3$

e) $\left(8a^3b^6\right)^{\frac{1}{3}}$

3 Write each of these in radical form and then evaluate.

a) $49^{\frac{1}{2}}$

b) $16^{\frac{1}{4}}$

c) $16^{\frac{3}{4}}$

d) $32^{\frac{4}{5}}$

e) $27^{\frac{2}{3}}$

f) $1000^{\frac{2}{3}}$

g) $4^{1.5}$

h) $81^{0.25}$

i) $9^{-0.5}$

j) $32^{-0.8}$

k) $\left(\dfrac{1}{8}\right)^{-\frac{5}{3}}$

l) $25^{-2.5}$

4 Use the rules of indices to simplify each of these. Write the answers in radical form.

a) $\sqrt{2} \times \sqrt{7}$

b) $\sqrt[5]{4} \times \sqrt[5]{128}$

c) $\sqrt[4]{8} \times \sqrt[4]{16}$

d) $\sqrt[3]{81} \times \sqrt[4]{27}$

e) $\sqrt[3]{25} \div \sqrt[5]{5}$

f) $\sqrt{27} \div \sqrt[12]{81}$

5 Use the rules of indices to simplify each of these. Write the answers with rational indices.

a) $d^{\frac{1}{3}} \times d^{\frac{2}{5}}$

b) $a^{\frac{4}{5}} \times a^{\frac{5}{6}}$

c) $g^{\frac{6}{7}} \div g^{\frac{3}{14}}$

d) $\left(s^{\frac{1}{3}}\right)^{\frac{3}{5}}$

e) $b^{\frac{2}{3}} \times b^{\frac{1}{2}} \div b^{\frac{1}{4}}$

f) $a^{\frac{1}{2}}b^{\frac{2}{3}} \times a^{\frac{2}{3}}b^{\frac{1}{4}}$

g) $c^{\frac{2}{3}}d^{\frac{5}{6}} \times c^{\frac{1}{2}}d \div (cd)^{\frac{1}{3}}$

6 Use the rules of indices to simplify each of these. Write the answers in radical form.

a) $a^{\frac{1}{2}} \times a^{\frac{2}{3}}$

b) $x^{\frac{1}{4}} \div x^{-\frac{2}{3}}$

c) $\left(c^{\frac{1}{2}}d^{\frac{1}{3}}\right)^5$

d) $(a^3b^2)^{\frac{1}{2}} \times (b^3a^4)^{-\frac{1}{3}}$

e) $x^{\frac{1}{5}}y^{\frac{1}{2}} \div x^{-\frac{1}{10}}y^{-1}$

f) $(c^{\frac{1}{4}}d^3)^{-\frac{1}{2}} \div (c^{\frac{1}{3}}d^{-\frac{1}{4}})^{-5}$

G Using the laws of indices to solve equations

We solve equations involving indices in the same way as other equations – by always doing the same thing to both sides of the equation. Take them one step at a time and you will find they are not as difficult as they look. Here are some tips to help you:

- Write as many of the numbers using the same base as you can.
- Write any bigger numbers in index form if possible.
- If the unknown variable is part of the index (or indices), change the equation so that the base on each side is the same. Then you can say that the indices on each side are also equal.

Examples

a) Solve the equation $a^2 = 256$ using the laws of indices.

$(a^2)^{\frac{1}{2}} = 256^{\frac{1}{2}}$ First take the square root of both sides (multiply by an index of $\frac{1}{2}$).

$a = (2^8)^{\frac{1}{2}}$ Now write 256 in index form and simplify.

$ = 2^{8 \times \frac{1}{2}}$

$ = 16$

b) Use the laws of indices to find the value of the variable in each of these equations.

i) $x^4 = 16$

$x^4 = 16$

$x^4 = 2^4$

$x = 2$

ii) $a^{\frac{1}{3}} = 3$

$a^{\frac{1}{3}} = 3$

$\left(a^{\frac{1}{3}}\right)^3 = 3^3$

$a^{\frac{1}{3} \times 3} = 27$

$a = 27$

iii) $4^x = 32$

$4^x = 32$

$(2^2)^x = 2^5$

$2^{2x} = 2^5$

$2x = 5$

$x = \frac{5}{2}$

$ = 2\frac{1}{2}$

iv) $7^d = \sqrt{343}$

$7^d = \sqrt{343}$

$7^d = \sqrt{7^3}$

$7^d = 7^{\frac{3}{2}}$

$d = \frac{3}{2}$

$ = 1\frac{1}{2}$

v) $4^n = \frac{1}{8}$

$$4^n = \frac{1}{8}$$
$$(2^2)^n = \frac{1}{2^3}$$
$$2^{2n} = 2^{-3}$$
$$2n = -3$$
$$n = -\frac{3}{2}$$
$$= -1\frac{1}{2}$$

c) Use the laws of indices to find the value of the variable in each of these equations.

i) $2^3 \times 4^2 = 2^v$

$$2^3 \times 4^2 = 2^v$$
$$2^3 \times (2^2)^2 = 2^v$$
$$2^3 \times 2^4 = 2^v$$
$$2^7 = 2^v$$
$$v = 7$$

ii) $3^{15} \div 27 = 3^w$

$$3^{15} \div 27 = 3^w$$
$$3^{15} \div 3^3 = 3^w$$
$$3^{15-3} = 3^w$$
$$3^{12} = 3^w$$
$$w = 12$$

iii) $8^{x+1} = 16^{x-3}$

$$8^{x+1} = 16^{x-3}$$
$$(2^3)^{x+1} = (2^4)^{x-3}$$
$$2^{3(x+1)} = 2^{4(x-3)}$$
$$3(x + 1) = 4(x - 3)$$
$$3x + 3 = 4x - 12$$
$$x = 15$$

Exercise 7

Use the laws of indices to find the value of the variable in each of these equations.

1 $3^a = 81$

2 $5^b = 125$

3 $9^c = 243$

4 $d^3 = 343$

5 $e^4 = 256$

6 $3f^2 = 48$

7 $5g^3 = 320$

8 $5^{-b} = 1$

9 $2^5 \div 8 = 2^x$

10 $2^x = \frac{1}{32}$

11 $3^k = \frac{1}{27}$

12 $5^x = \frac{1}{625}$

13 $(4^5)^2 = 2^m$

14 $n^{\frac{1}{2}} = 4$

15 $x^{\frac{1}{5}} = 2$

16 $3^p = \sqrt{27}$

17 $9^x = \sqrt{243}$

18 $\sqrt{4^r} = \frac{1}{8}$

19 $9^3 \times 27 = 3^s$

20 $8^3 \div 2^t = 4$

21 $9^{2x-1} = 27^{2x-5}$

Algebraic multiplication and division

Key vocabulary

distributive law	expand	polynomial
dividend	monomial	quotient
divisor	perfect square	

A Expansion using the distributive law

Revision

When we expand brackets we multiply them out using the distributive law.

For example:

$2 \times (5 + 3) = 2 \times 5 + 2 \times 3$
$= 10 + 6$
$= 16$

We can also use the **distributive law** to remove the brackets from an expression in algebra.

For example:

$2(3a + 4b) = 2 \times 3a + 2 \times 4b$
$= 6a + 8b$

Remember that the rules for multiplying in algebra are the same as the rules for multiplying numbers.

$(+) \times (+) = (+)$	$(+2)(+3) = 6$	$(+a)(+b) = ab$
$(-) \times (-) = (+)$	$(-2)(-3) = 6$	$(-a)(-b) = ab$
$(+) \times (-) = (-)$	$(+2)(-3) = -6$	$(+a)(-b) = -ab$
$(-) \times (+) = (-)$	$(-2)(+3) = -6$	$(-a)(+b) = -ab$

So, in general, the distributive law can be written as
$a(b + c) = ab + ac$.

Examples

Expand each of these expressions.

a) $2(x + 1)$

$$2(x + 1) = 2 \times x + 2 \times 1$$
$$= 2x + 2$$

b) $4(3x - 5)$

$$4(3x - 5) = 4 \times 3x + 4 \times (-5)$$
$$= 12x - 20$$

Remember, if the number in front of a bracket is negative, the sign of **each term inside** the bracket is changed when the bracket is removed.

Examples

Expand each of these expressions.

a) $-2(a + 5)$

$$-2(a + 5) = -2 \times a - 2 \times 5$$
$$= -2a - 10$$

b) $-3(2b - 7)$

$$-3(2b - 7) = -3 \times 2b - 3 \times (-7)$$
$$= -6b + 21$$

Sometimes the value outside the bracket is a variable.

Examples

Expand each of these expressions.

a) $a(a + 3)$

$$a(a + 3) = a \times a + a \times 3$$
$$= a^2 + 3a$$

b) $b(2b - 3)$

$$b(2b - 3) = b \times 2b + b \times (-3)$$
$$= 2b^2 - 3b$$

c) $-c(c + 4)$

$$-c(c + 4) = -c \times c - c \times 4$$
$$= -c^2 - 4c$$

d) $(2d + 1)d$

$$(2d + 1)d = 2d \times d + 1 \times d$$
$$= 2d^2 + d$$

When we simplify an expression involving two brackets, we expand each bracket separately, then collect like terms.

Examples

Expand and simplify each of these expressions.

a) $4(2a - 3) + 3(a + 2)$

$4(2a - 3) + 3(a + 2) = 8a - 12 + 3a + 6 = 11a - 6$

b) $2(a + 3b) - 3(b - a)$

$2(a + 3b) - 3(b - a) = 2a + 6b - 3b + 3a = 5a + 3b$

Check that you are confident with this revision work before going on with this unit. Make sure you know how to answer all the questions in Exercise 7 in Unit 9 of Coursebook 1.

B Expansion of two expressions in the form $(a + b)(c + d)$

If we have two expressions to multiply together, for example $(a + b)(c + d)$, **each term** in the first expression must be multiplied by **each term** in the second expression.

We can make this process a bit easier to understand if we replace one of the expressions by a single letter, say k, to start with.

So, let $(c + d) = k$

Then $\quad\quad\quad (a + b)(c + d) = (a + b)k$

$= ak + bk$

Now substitute $k = (c + d)$ $\quad = a(c + d) + b(c + d)$

$= ac + ad + bc + bd$

We can get the same answer from the two expressions by expanding directly.

$(a + b)(c + d) = ac + ad + bc + bd$

Any two expressions, each with **two terms**, can be expanded fully if you follow the four separate multiplications as shown.

Sometimes these four multiplications are given names.

1 = **First** terms
2 = **Outer** terms
3 = **Inner** terms
4 = **Last** terms

So the order is First terms, Outer terms, Inner terms, Last terms. You can remember this order with the word **FOIL**!

Examples

a) Expand each of these expressions and simplify.

   i) $(x + 2)(x + 5)$

$$(x + 2)(x + 5) = x^2 + 5x + 2x + 10$$
$$= x^2 + 7x + 10$$

   ii) $(x + 3)(x - 5)$

$$(x + 3)(x - 5) = x^2 - 5x + 3x - 15$$
$$= x^2 - 2x - 15$$

   iii) $(x - 2)(x + 3)$

$$(x - 2)(x + 3) = x^2 + 3x - 2x - 6$$
$$= x^2 + x - 6$$

   iv) $(x - 4)(x - 6)$

$$(x - 4)(x - 6) = x^2 - 6x - 4x + 24$$
$$= x^2 - 10x + 24$$

b) Expand each of these expressions and simplify.

   i) $(2a + 3)(a + 4)$

$$(2a + 3)(a + 4) = 2a^2 + 8a + 3a + 12$$
$$= 2a^2 + 11a + 12$$

   ii) $(3b + 2)(b - 3)$

$$(3b + 2)(b - 3) = 3b^2 - 9b + 2b - 6$$
$$= 3b^2 - 7b - 6$$

   iii) $(2c - 3)(4c + 5)$

$$(2c - 3)(4c + 5) = 8c^2 + 10c - 12c - 15$$
$$= 8c^2 - 2c - 15$$

   iv) $(d - 2)(3d - 4)$

$$(d - 2)(3d - 4) = 3d^2 - 4d - 6d + 8$$
$$= 3d^2 - 10d + 8$$

Exercise 1

Expand and simplify each of these expressions.

1 $(3a + 2)(a + 4)$ 2 $(2c + 3)(c + 5)$ 3 $(1 + x)(9 + 5x)$

4 $(5y - 3)(y - 4)$ 5 $(z - 3)(2z - 7)$ 6 $(p - 2)(4p - 5)$

7 $(c + 5)(3c - 1)$　　　　**8** $(5b - 2)(b + 3)$　　　　**9** $(3a - 2)(2a + 1)$

10 $(5a - 3)(3a - 4)$　　**11** $(4a + 3)(3a - 4)$　　**12** $(4b - 1)(2 - 3b)$

13 $(3p - 2)(3p - 5)$　　**14** $(4r - 5)(3 - 5r)$　　**15** $(5 - 2t)(3 + t)$

16 $(x - 7)(x + 8) + (x + 2)(x - 3)$

17 $(x + 2y)(x - 6y) - (x - 3y)(x - y)$

18 $(2x - 5y)(3x + y) + 5y(4x + y)$

19 $5p^2 - 4(p + q)(2p - 3q)$

C Expansion of two expressions in the form $(a + b)(c + d + e)$

It doesn't matter how many terms we have in each of the expressions – we use the same rules to multiply **any** two expressions:

> **Every term** in one expression must be multiplied by **every term** in the other expression. We can then simplify by combining like terms.

We expand the two expressions $(a + b)(c + d + e)$ like this:

$$(a + b)(c + d + e) = ac + ad + ae + bc + bd + be$$

Examples　　Expand each of these expressions and simplify.

a) $(2x - y)(3x + 2y + 1)$

$(2x - y)(3x + 2y + 1) = 6x^2 + 4xy + 2x - 3xy - 2y^2 - y$
$\qquad\qquad\qquad\qquad = 6x^2 + xy + 2x - 2y^2 - y$

b) $(5x - 4)(2x^2 + 3x + 2)$

$(5x - 4)(2x^2 + 3x + 2) = 10x^3 + 15x^2 + 10x - 8x^2 - 12x - 8$
$\qquad\qquad\qquad\qquad = 10x^3 + 7x^2 - 2x - 8$

c) $(3a - 4)(2a^3 - 7a + 6)$

$(3a - 4)(2a^3 - 7a + 6) = 6a^4 - 21a^2 + 18a - 8a^3 + 28a - 24$
$\qquad\qquad\qquad\qquad = 6a^4 - 8a^3 - 21a^2 + 46a - 24$

d) $(d^2 + 3)(3d^2 - 4d - 1)$

$(d^2 + 3)(3d^2 - 4d - 1) = 3d^4 - 4d^3 - d^2 + 9d^2 - 12d - 3$
$\qquad\qquad\qquad\qquad = 3d^4 - 4d^3 + 8d^2 - 12d - 3$

Exercise 2

Expand and simplify each of these expressions.

1 $(x + 2)(x^2 + x + 1)$ 2 $(x + 1)(x^2 - x - 1)$ 3 $(x - 1)(x^2 + 2x - 1)$

4 $(x - 2)(x^2 - 2x + 3)$ 5 $(a + 3)(a^2 + 3a - 2)$ 6 $(a - 3)(a^2 - 3a + 4)$

7 $(2a - 1)(a^2 + 4a + 1)$ 8 $(3a + 2)(a^2 - 4a + 2)$

D Some special products of expansion

Expansion when the expressions in the brackets are the same, $(a + b)(a + b)$

We know that any number multiplied by itself gives the **square** of that number.

This is also true for any expression in a bracket, such as $(a + b)$. If we multiply this by itself we get a **square**: $(a + b) \times (a + b) = (a + b)^2$.

In Unit 6 we defined a perfect square as the number made by multiplying a **whole number** by itself.

In algebra, a **perfect square** is made by multiplying the **whole of an expression** by itself.

So what does a perfect square in algebra look like?

$$(a + b)(a + b) = a^2 + ab + ba + b^2$$
$$= a^2 + 2ab + b^2 \quad \text{(Remember, } ab = ba.)$$

$a^2 + \underline{2ab} + b^2$ is a **perfect square**.

$$(a - b)(a - b) = a^2 - ab - ba + b^2$$
$$= a^2 - 2ab + b^2$$

$a^2 - 2ab + b^2$ also is a **perfect square**.

There are two things to notice about perfect squares in algebra:

● The term b^2 in both products is **positive**. This is because the signs of both bs in the original expressions are **the same** [remember, $(+) \times (+) = (+)$ and $(-) \times (-) = (+)$].

● The term $2ab$ in each product has the **same sign** as the **original** expression.
$(a + b)(a + b)$ gives a middle term of $+ 2ab$.
$(a - b)(a - b)$ gives a middle term of $- 2ab$.

Examples   Expand these expressions.

a) $(3u + 4)^2$

General: $(a\ \ +\ \ b)^2\ =\ \ a^2\ \ +\ \ 2ab\ \ +\ \ b^2$
$\qquad\qquad\downarrow\qquad\downarrow\qquad\quad\downarrow\qquad\quad\downarrow\qquad\quad\downarrow$
So: $(3u\ +\ 4)^2\ =\ (3u)^2\ +\ 2(3u)(4)\ +\ 4^2$
(Replace a with $3u$ and b with 4.)
$$= 9u^2 + 24u + 16$$

b) $(2v - 3)^2$

General: $(a\ \ -\ \ b)^2\ =\ \ a^2\ \ -\ \ 2ab\ \ +\ \ b^2$
$\qquad\qquad\downarrow\qquad\downarrow\qquad\quad\downarrow\qquad\quad\downarrow\qquad\quad\downarrow$
So: $(2v\ -\ 3)^2\ =\ (2v)^2\ -\ 2(2v)(3)\ +\ 3^2$
(Replace a with $2v$ and b with 3.)
$$= 4v^2 - 12v + 9$$

c) $(4x + y)^2$

$(4x + y)^2 = (4x)^2 + 2(4x)(y) + (y)^2$
$$= 16x^2 + 8xy + y^2$$

d) $(3m - n)^2$

$(3m - n)^2 = (3m)^2 - 2(3m)(n) + (n)^2$
$$= 9m^2 - 6mn + n^2$$

We could multiply the two expressions the long way, using FOIL – we would get the same answer. However, it is quicker to use the shortcuts demonstrated above, and being able to recognise the pattern of a perfect square will help later on, when we work backwards to find the factors of an expression. So make sure you use the shortcuts when you do Exercise 3.

Exercise 3

Expand and simplify each of these expressions.

1 $(x + 1)^2$	2 $(x + 5)^2$	3 $(x + 8)^2$	4 $(x + 10)^2$
5 $(x + 11)^2$	6 $(x - 3)^2$	7 $(x - 5)^2$	8 $(x - 7)^2$
9 $(x - 4)^2$	10 $(x - 8)^2$	11 $(x - 6)^2$	12 $(x - 12)^2$
13 $(x + y)^2$	14 $(m + n)^2$	15 $(p + q)^2$	16 $(s + t)^2$
17 $(s - t)^2$	18 $(p - q)^2$	19 $(2x + 1)^2$	20 $(3a + 1)^2$

21 $(3m + 2)^2$ 22 $(2a + 3)^2$ 23 $(5a - 2)^2$ 24 $(4a - 3)^2$

25 $(a - 3b)^2 - (3a - b)^2$ 26 $4(a + 3)^2 + 3(a - 4)(a + 7)$

27 $2(2a + 1)(3a - 2) - 3(4a - 1)^2$ 28 $2p(5p - 4q) - (3p + 2q)^2$

Expansion when the expressions in the brackets have the same numbers and variables but opposite signs, $(a + b)(a - b)$

If we expand an expression of the form $(a + b)(a - b)$ using the distributive law, we get

$$(a + b)(a - b) = a^2 - ab + ba - b^2$$
$$= a^2 - b^2 \quad (-ab \text{ and } +ab \text{ cancel each other out.})$$

Because a^2 is a perfect square (we know that $\sqrt{a^2} = a$) and b^2 is a perfect square ($\sqrt{b^2} = b$), we can say that $a^2 - b^2$ is the **difference of perfect squares**.

$$(a + b)(a - b) = a^2 - b^2$$

This is a very useful rule that helps us to expand brackets with the same expressions but opposite signs quickly without having to use the full process of the distributive law every time.

Examples

Expand these expressions.

a) $(x + 4)(x - 4)$

$(x + 4)(x - 4) = x^2 - 16$

b) $(x - 3)(x + 3)$

$(x - 3)(x + 3) = x^2 - 9$

c) $(k + 7)(k - 7)$

$(k + 7)(k - 7) = k^2 - 49$

d) $(2p - 3)(2p + 3)$

$(2p - 3)(2p + 3) = 4p^2 - 9$

NOTE: Part **d)** of the example above shows that it is important to square the whole of the first term (the a term). Here $(2p)^2 = 4p^2$. Squaring just the p, and forgetting about the 2, would give $2p^2$, which is incorrect.

Again, we could multiply the two expressions the long way, using FOIL, but the shortcuts are quicker and it is useful to be able to recognise the pattern that makes the difference of perfect squares. Make sure you use the shortcuts when you do Exercise 4.

Exercise 4

Expand and simplify each of these expressions.

1 $(x + 1)(x - 1)$ 2 $(y + 2)(y - 2)$ 3 $(m + 10)(m - 10)$

4 $(a + 5)(a - 5)$ 5 $(x + 8)(x - 8)$ 6 $(p - 6)(p + 6)$

7 $(b - 11)(b + 11)$ 8 $(5 - b)(5 + b)$ 9 $(8 - m)(8 + m)$

10 $(r - s)(r + s)$ 11 $(p + q)(p - q)$ 12 $(m - n)(m + n)$

13 $(2p - 1)(2p + 1)$ 14 $(4q - 1)(4q + 1)$ 15 $(3a + 4)(3a - 4)$

16 $(5b - 3)(5b + 3)$ 17 $(e + d)(d - e)$ 18 $(7g - 2)(2 + 7g)$

E Using the rules of algebra to make calculations easier

We can apply these rules for algebraic expansions to numerical calculations.

Examples

Use the rules for algebraic expansions to find answers to these calculations.

a) 201×199

$$
\begin{aligned}
201 \times 199 &= (200 + 1)(200 - 1) \\
&= 200^2 - 1^2 \\
&= 40\,000 - 1 \\
&= 39\,999
\end{aligned}
$$
(Difference of two perfect squares)

b) $(797)^2$

$$
\begin{aligned}
(797)^2 &= (800 - 3)^2 \\
&= (800)^2 - 2(800)(3) + 3^2 \\
&= 640\,000 - 4800 + 9 \\
&= 635\,209
\end{aligned}
$$
(An algebraic perfect square)

c) $(39)^2 + 78 + 1$

$$\begin{aligned}(39)^2 + 78 + 1 &= (39)^2 + 2(39)(1) + 1^2 \\ &= (39 + 1)^2 \\ &= (39 + 1)^2 \\ &= 40^2 \\ &= 1600\end{aligned}$$

(In the form of an algebraic square)

Exercise 5

Use the rules of algebraic expansion to help you work out answers to these calculations.

1 502×498

2 305×295

3 98×102

4 $(9\,001)^2$

5 $(699)^2$

6 $(892)^2$

7 $(69)^2 + 138 + 1$

8 $(78)^2 + 312 + 4$

9 $(301)^2 - 602 + 1$

F Division of algebraic expressions

An algebraic expression with **only one term** is called a monomial. An algebraic expression with **more than one term** is called a polynomial.

Revision

Dividing one monomial by another monomial

We have already learned how to divide one **monomial** by another **monomial**.

To divide a monomial with a **variable term** by a monomial with a **constant term**, we simply divide the number parts as normal, and leave the variable letters the same to make the new term.

Examples

a) $8a \div 2 = 4a$
b) $15p \div (-5) = -3p$
c) $(-9x) \div 3 = -3x$
d) $(-12y^2) \div (-4) = +3y^2$

To divide **two monomials**, each with **variable terms**, we divide the coefficients of each term as normal and use the index rules to divide any powers of the same letter.

Examples

a) $9x \div x = 9$
b) $6z \div (-2z) = -3$
c) $(-8a^2) \div 2a = -4a$
d) $(-15d^4) \div (-3d^2) = +5d^2$

Dividing a polynomial by a monomial

To divide **any polynomial expression** by a **monomial**, we simply divide each term in the expression by the term in the monomial.

Examples

a) Divide $9a^5 - 6a^2$ by $3a^2$.

$$\frac{9a^5 - 6a^2}{3a^2} = \frac{9a^5}{3a^2} - \frac{6a^2}{3a^2}$$

$$= \frac{{}^{3}\cancel{9}a^{5\,3}}{\cancel{3}\cancel{a^2}} - \frac{{}^{2}\cancel{6}\cancel{a^2}}{\cancel{3}\cancel{a^2}}$$

$$= 3a^3 - 2$$

b) Divide $12x^3y + 20x^2y^3$ by $4xy^2$.

$$\frac{12x^3y + 20x^2y^3}{4xy^2} = \frac{12x^3y}{4xy^2} + \frac{20x^2y^3}{4xy^2}$$

$$= \frac{{}^{3}\cancel{12}x^{3\,2}\cancel{y}}{\cancel{4}\cancel{x}y^{2\,1}} + \frac{{}^{5}\cancel{20}x^{2\,1}y^{3\,1}}{\cancel{4}\cancel{x}\cancel{y^2}}$$

$$= \frac{3x^2}{y} + 5xy$$

Exercise 6

1 Divide these monomials.

a) $-70a^3 \div 14a^2$
b) $63a^4b^5c^6 \div -9a^2b^4c^3$
c) $24x^3y^3 \div -8y^2$
d) $x^{10} \div (-x)^7$
e) $15a^4b \div -5a^3b$
f) $\frac{1}{4}x^5y^3 \div -\frac{3}{2}x^4y$
g) $-24x^4d^3 \div -2x^2d^5$
h) $-\frac{3}{4}a^2bc^3 \div -\frac{2}{9}abc^2$
i) $-6g^3h \div 2g^3h^2$

2 Divide these polynomials by the monomials as shown.

a) $\dfrac{8x - 10y + 6c}{2}$
b) $\dfrac{3ax - 6bx - 15x}{-3x}$
c) $\dfrac{m^3n - 2m^2n^2 + mn^3}{mn}$

d) $\dfrac{-4a^2b^3 - 8ab^2 + 6ab}{-2ab}$

e) $\dfrac{9x^3y - 15x^2y^2 - 6x^4y^3}{3x^2y}$

f) $\dfrac{15a^3b^4 - 10a^4b^3 - 25a^3b^6}{-5a^3b^2}$

g) $\dfrac{-14x^6y^3 - 21x^4y^5 + 7x^5y^4}{7x^2y^2}$

h) $\dfrac{3a^2b + 4ab^3 - 7a^5b^2}{-2a^2b^2}$

i) $\dfrac{-\frac{1}{2}m^6n^6 + \frac{2}{5}m^7n^7 - \frac{4}{5}m^8n^8}{-\frac{1}{10}m^5n^3}$

j) $\dfrac{-\frac{5}{7}a^2x^3 - a^3x^2 + \frac{2}{3}a^5x^4}{-\frac{3}{7}ax}$

k) $\dfrac{x^2y^3 - \frac{11}{16}x^3y^2 + \frac{5}{8}x^3y^3}{-\frac{1}{4}x^2y^2}$

Dividing one polynomial by another polynomial

We have already learned to use long division to divide bigger numbers.

For example, we could calculate $156 \div 13$ like this.

$$
\begin{array}{r}
12 \\
13\overline{)156} \\
-\ 13 \\
\hline
26 \\
-\ \ 26 \\
\hline
00
\end{array}
$$

Alternatively, we could write the numbers differently, to show the place value of each digit.

$156 = 100 + 50 + 6$ and $13 = 10 + 3$

Writing the numbers in this way, the long division calculation would look like this.

$$
\begin{array}{r}
10 + 2 \\
10 + 3\overline{)100 + 50 + 6} \\
\underline{100 + 30} \qquad = 10 \times (10 + 3) \\
20 + 6 \\
\underline{20 + 6} \qquad = \ \ 2 \times (10 + 3) \\
0
\end{array}
$$

If we replace 10 by an x, $100 = 10 \times 10$ becomes $x \times x = x^2$

$\qquad\qquad\qquad\qquad\qquad 50 = 5 \times 10$ becomes $5 \times x = 5x$

$\qquad\qquad\qquad\qquad\qquad 10 + 3$ \qquad becomes $x + 3$.

Now the long division calculation looks like this.

$$
\begin{array}{r}
x + 2 \\
x + 3 \overline{) x^2 + 5x + 6} \\
\end{array}
$$

$$x^2 + 3x \qquad = x \times (x + 3)$$

$$2x + 6$$

$$2x + 6 \quad = 2 \times (x + 3)$$

$$0$$

By looking at this calculation, we can work out the steps we need to follow **to divide one polynomial by another polynomial.**

1 Arrange the terms in both polynomials in descending order of the powers of the variables.

2 Divide the first term of the polynomial to be divided (called the dividend) by the first term in the polynomial you are dividing by (called the divisor) to find the first term of the answer (called the quotient).

3 Multiply the **whole** of the divisor polynomial by this first term in the quotient, and subtract this answer from the dividend polynomial.

4 Using the remainder as the new dividend polynomial, repeat steps 2 and 3. Continue for as long as possible.

Examples

Work out these algebraic division calculations using long division.

a) $(2x^2 + 11x + 14) \div (x + 2)$

$$
\begin{array}{r}
2x + 7 \\
x + 2 \overline{) 2x^2 + 11x + 14} \\
\end{array}
$$

$$2x^2 + 4x$$

$$7x + 14$$

$$7x + 14$$

$$0$$

So $(2x^2 + 11x + 14) \div (x + 2) = 2x + 7$.

b) $(8x^2 + 18xy - 45y^2) \div (2x - 3y)$

$$
\begin{array}{r}
4x + 15y \\
2x - 3y \overline{) 8x^2 + 18xy - 45y^2} \\
\end{array}
$$

$$8x^2 - 12xy$$

$$+30xy - 45y^2 \quad [18xy - (-12xy) = 30xy]$$

$$30xy - 45y^2$$

$$0$$

So $(8x^2 + 18xy - 45y^2) \div (2x - 3y) = 4x + 15y$

Exercise 7

Divide these polynomials by using long division.

1 $(x^2 + 7x + 12) \div (x + 4)$

2 $(2x^2 - 13x + 21) \div (x - 3)$

3 $(6x^2 - 5x + 1) \div (2x - 1)$

4 $(a^2 + 3a - 54) \div (a - 6)$

5 $(12x^2 + 7xy - 12y^2) \div (3x + 4y)$

6 $(3a^3 + 3a^2 - 15a - 15) \div (3a + 3)$

7 $(8a^2 - 14ab - 15b^2) \div (2a - 5b)$

8 $(6x^3 + x^2y - 16xy^2 - 6y^3) \div (2x + 3y)$

9 $(4x^2 + \langle 0xy \rangle - 9y^2) \div (2x + 3y)$

10 $(x^2 + \langle 0xy \rangle - 16y^2) \div (x + 4y)$

11 $(x^6 + \langle 0x^4 \rangle + \langle 0x^2 \rangle - 8) \div (x^2 - 2)$

12 $(6x^3 - 13x^2 - 13x + 30) \div (2x^2 - x - 6)$

Factorising quadratic expressions

Key vocabulary

expand   factorise   quadratic

A Expansion of algebraic expressions

Revision

Expansion of two brackets in the form $(a + b)(c + d)$

Any two brackets can be expanded fully if you follow the four separate multiplications as shown.

$$(a + b)(c + d) = ac + ad + bc + bd$$

Expansion of two brackets in the form $(a + b)(a + b)$ or $(a - b)(a - b)$

If the expressions in the brackets are exactly the same, the product expression is called a **perfect square**.

$(a + b)(a + b) = a^2 + 2ab + b^2$
$(a - b)(a - b) = a^2 - 2ab + b^2$

Remember:

- The term b^2 in both products is **positive**.
- The term $2ab$ in each product has the **same sign** as the **original** expression.

Expansion of two brackets in the form $(a + b)(a - b)$

If the expressions in the brackets have the same numbers and variables but opposite signs, the product expression is called a **difference of perfect squares**.

$(a + b)(a - b) = a^2 - b^2$

Check that you are confident with this revision work before going on with this unit. Look at Exercises 1, 2, 3 and 4 of Unit 8 and make sure you know how to answer all the questions.

B Factorisation of quadratic expressions

We have already learned how to expand expressions – now let's try to do it all in reverse.

Factorisation is the reverse of **expansion**.

In a quadratic expression the highest power of the variable is 2.

Finding the highest common factor (HCF) of an expression

We already know how to factorise numbers into prime factors.

If we remember that a^2 is the same as $a \times a$, then we can write out all the factors for any term.

Examples   Find the highest common factor (HCF) of each of these pairs of terms.

a) $3a$ and $3b$

$3a = \boxed{3} \times a$
$3b = \boxed{3} \times b$
So the HCF of $3a$ and $3b$ is 3.

b) $5a$ and 10

$5a = \boxed{5} \times a$
$10 = 2 \times \boxed{5}$
So the HCF of $5a$ and 10 is 5.

c) p^2 and $6p$

$p^2 = \boxed{p} \times p$
$6p = 2 \times 3 \times \boxed{p}$
So the HCF of p^2 and $6p$ is p.

d) $4m^2$ and $10m$

$4m^2 = \boxed{2} \times 2 \times \boxed{m} \times m$
$10m = \boxed{2} \times 5 \times \boxed{m}$
So the HCF of $4m^2$ and $10m$ is $2m$.

e) a^2b^2 and $3ab$

$a^2b^2 = \boxed{a} \times a \times \boxed{b} \times b$
$3ab = 3 \times \boxed{a} \times \boxed{b}$
So the HCF of a^2b^2 and $3ab$ is ab.

f) $3x^2y$ and $12y^2x$

$3x^2y = \textcircled{3} \times \textcircled{x} \times x \times \textcircled{y}$

$12y^2x = 2 \times 2 \times \textcircled{3} \times \textcircled{y} \times y \times \textcircled{x}$

So the HCF of $3x^2y$ and $12y^2x$ is $3xy$.

 **Exercise 1**

Find the highest common factor (HCF) for each of these pairs of expressions.

1 $2x$ and x 2 $12a$ and 6 3 $24m$ and $12n$

4 $2x$ and $4x$ 5 $14m$ and $7m$ 6 $21r$ and $9r$

7 a^2 and a 8 $2y^2$ and $2y$ 9 $6d^2$ and $12d$

10 ab and ab^2 11 ab and a^2b 12 mn and m^2n^2

13 $6a^2b$ and $12ab^2$ 14 $4d^3e^2$ and $6d^2e^3$ 15 $-28cde$ and $-42ce^2f$

Factorising expressions by taking out the HCF

In **Unit 8** we learned how to **expand** brackets, using the **distributive law**: $a(b + c) = ab + ac$.

In reverse, $ab + ac = a(b + c)$. This procedure involves taking out the highest common factor.

In the expression $ab + ac$, the highest common factor is a. This is taken out and placed in front of the brackets. We make the expression inside the brackets by dividing each term of the original expression by the HCF.

> Writing an algebraic expression as the product of its factors is called **factorisation**.

> The factorised form of an expression is the **simplest form** of that expression.

Examples Factorise each of these expressions by taking out the highest common factor (HCF).

a) $4x + 8$

$$4x + 8 = 4 \times x + 4 \times 2$$
$$= 4(x + 2)$$

b) $x^2 - x$

$$x^2 - x = x \times x - x \times 1$$
$$= x(x - 1)$$

c) $6a + 9$

$$6a + 9 = 3 \times 2a + 3 \times 3$$
$$= 3(2a + 3)$$

d) $-3m - 12$

$$-3m - 12 = (-3 \times m) + (-3 \times 4)$$
$$= -3(m + 4)$$

e) $2a + 4b - 8c$

$$2a + 4b - 8c = 2 \times a + 2 \times 2b - 2 \times 4c$$
$$= 2(a + 2b - 4c)$$

f) $9x^2 + 24xy - 12x$

$$9x^2 + 24xy - 12x = 3x \times 3x + 3x \times 8y + 3x \times (-4)$$
$$= 3x(3x + 8y - 4)$$

g) $x(x + 2) + 3(x + 2)$

We can see that $(x + 2)$ is common to both $x(x + 2)$ and $3(x + 2)$ so it can be taken out.
$$x(x + 2) + 3(x + 2) = (x + 2)(x + 3)$$

h) $k(2k - 3) + (2k - 3)$

We see that $(2k - 3)$ is common to both terms.
$$k(2k - 3) + (2k - 3) = (2k - 3)(k + 1)$$

Exercise 2

1 Factorise each of these expressions completely by taking out the HCF.

a) $5x + 15$ b) $7x + 28$ c) $15x - 25$
d) $24x - 8$ e) $32x + 48$ f) $3x - 9x^2$
g) $5x^2 - 10x$ h) $xy - y^2$ i) $x^3 - x^2y$
j) $3x^2 - 6xy$ k) $4x^2 - 16x^2y$ l) $15a + 45a^2$
m) $9x^2y^2 - 3x^2z^2$ n) $4x^3 + x^2 - x$ o) $6a^3 - 4a^2 - 2a$
p) $12xy + 9x^2y + 3x^3$ q) $3x^4y - 6x^3y + 9x^2y^3$ r) $7a^3 - 7a^2b + 14ab^2$
s) $2a^2b^3 - 6a^2b^2 + 2a^3b$ t) $2a^5 - 6a^4b + 2a^3b^2$

2 Factorise each of these expressions completely by taking out the HCF.

a) $x(x + 1) + 2(x + 1)$
b) $t(t - 2) + 2(t - 2)$
c) $y(y + 3) - 3(y + 3)$
d) $s(s + 6) - 6(s + 6)$
e) $b(4b - 3) - 3(4b - 3)$
f) $y(4y + 1) - 2(4y + 1)$
g) $x(1 - 3x) + 2(1 - 3x)$
h) $y(y + 1) + (y + 1)$
i) $m(2m + 3) + (2m + 3)$
j) $t(3t + 5) + (3t + 5)$
k) $q(q - 4) - 4(q - 4)$
l) $x(x + y) - y(x + y)$
m) $x(x - y) + y(x - y)$
n) $a(x + y) + b(x + y)$
o) $(a + b)^2 + 2(a + b)$
p) $(a - b)^2 - 5(a - b)$
q) $x(m + n) + y(m + n) + z(m + n)$
r) $a(x + y + z) + b(x + y + z)$

Factorising perfect squares

Do you remember what an algebraic **perfect square** looks like?

$(a + b)^2 = a^2 + 2ab + b^2$
$(a - b)^2 = a^2 - 2ab + b^2$

In reverse, $a^2 + 2ab + b^2 = (a + b)^2$ or $(a + b)(a + b)$
$a^2 - 2ab + b^2 = (a - b)^2$ or $(a - b)(a - b)$

So it is easy to **factorise** a perfect square. But first we must recognise that the expression is a perfect square. Here are some questions we can use to help us decide whether an expression is a perfect square.

- Is the first term a perfect square ($= a^2$)?
- Is the last term a perfect square ($= b^2$)?
- Does the middle term equal $2 \times a \times b$? In other words, does it equal $2 \times \sqrt{\text{first term}} \times \sqrt{\text{second term}}$?

If the answer to all these questions is 'Yes', then the expression is a **perfect square** and it will factorise following the pattern above.

Remember that the term $2ab$ in a perfect square has the **same sign** as the **original** expression.

- If it is $+2ab$ then the factors are $(a + b)(a + b)$.
- If it is $-2ab$ then the factors are $(a - b)(a - b)$.

Use these rules to factorise a perfect square:

$a^2 + 2ab + b^2 = (\sqrt{\text{first term}} + \sqrt{\text{last term}})(\sqrt{\text{first term}} + \sqrt{\text{last term}})$
$\quad or \ (\sqrt{\text{first term}} + \sqrt{\text{last term}})^2$

$a^2 - 2ab + b^2 = (\sqrt{\text{first term}} - \sqrt{\text{last term}})(\sqrt{\text{first term}} - \sqrt{\text{last term}})$
$\quad or \ (\sqrt{\text{first term}} - \sqrt{\text{last term}})^2$

Examples

Factorise these expressions.

a) $x^2 + 10x + 25$

The first term is a perfect square: $\sqrt{x^2} = x$
The last term is a perfect square: $\sqrt{25} = 5$
$2 \times \sqrt{\text{first term}} \times \sqrt{\text{last term}} = 2 \times x \times 5 = 10x$ and this is the same as the middle term.
The expression is a perfect square.

$$x^2 + 10x + 25 = \left(\sqrt{x^2} + \sqrt{25}\right)\left(\sqrt{x^2} + \sqrt{25}\right)$$
$$= (x + 5)(x + 5) \text{ or } (x + 5)^2$$

b) $4x^2 - 12x + 9$

The first term is a perfect square: $\sqrt{4x^2} = 2x$
The last term is a perfect square: $\sqrt{9} = 3$
$2 \times \sqrt{\text{first term}} \times \sqrt{\text{last term}} = 2 \times 2x \times 3 = 12x$ and this is the same as the middle term.
The expression is a perfect square. Notice that the middle term has a negative sign.

$$4x^2 - 12x + 9 = \left(\sqrt{4x^2} - \sqrt{9}\right)\left(\sqrt{4x^2} - \sqrt{9}\right)$$
$$= (2x - 3)(2x - 3) \text{ or } (2x - 3)^2$$

c) $x^2 + 14xy + 49y^2$

The first term is a perfect square: $\sqrt{x^2} = x$
The last term is a perfect square: $\sqrt{49y^2} = 7y$
$2 \times \sqrt{\text{first term}} \times \sqrt{\text{last term}} = 2 \times x \times 7y = 14xy$ and this is the same as the middle term.
The expression is a perfect square.

$$x^2 + 14xy + 49y^2 = \left(\sqrt{x^2} + \sqrt{49y^2}\right)\left(\sqrt{x^2} + \sqrt{49y^2}\right)$$
$$= (x + 7y)(x + 7y) \text{ or } (x + 7y)^2$$

d) $16x^2 - 16xy + 4y^2$

The first term is a perfect square: $\sqrt{16x^2} = 4x$
The last term is a perfect square: $\sqrt{4y^2} = 2y$
$2 \times \sqrt{\text{first term}} \times \sqrt{\text{last term}} = 2 \times 4x \times 2y = 16xy$ and this is the same as the middle term.
The expression is a perfect square. Notice that the middle term has a negative sign.

$$16x^2 - 16xy + 4y^2 = \left(\sqrt{16x^2} - \sqrt{4y^2}\right)\left(\sqrt{16x^2} - \sqrt{4y^2}\right)$$
$$= (4x - 2y)(4x - 2y) \text{ or } (4x - 2y)^2$$

Exercise 3

Factorise each of these expressions (check whether they are perfect squares first).

1 $a^2 + 8a + 16$ 2 $1 + 2x + x^2$ 3 $z^2 - 6z + 9$

4 $c^2 + 10c + 25$ 5 $4a^2 + 4a + 1$ 6 $9c^2 - 6c + 1$

7 $4 - 12b + 9b^2$ 8 $25y^2 - 30y + 9$ 9 $4z^2 - 28z + 49$

10 $9a^2 - 12a + 4$ 11 $25b^2 - 20b + 4$ 12 $16c^2 + 24c + 9$

13 $81 + 90x + 25x^2$ 14 $16p^2 - 40p + 25$

Factorising the difference of two perfect squares

We have already learned that $(a + b)(a - b) = a^2 - b^2$. The result is the difference of two perfect squares. In reverse:

$$a^2 - b^2 = (a + b)(a - b)$$

Examples

Factorise these expressions.

a) $x^2 - 4$

$$\begin{aligned} x^2 - 4 &= x^2 - 2^2 \\ &= (x - 2)(x + 2) \end{aligned}$$

b) $a^2 - 64$

$$\begin{aligned} a^2 - 64 &= a^2 - 8^2 \\ &= (a - 8)(a + 8) \end{aligned}$$

c) $4m^2 - 25$

$$\begin{aligned} 4m^2 - 25 &= (2m)^2 - 5^2 \\ &= (2m - 5)(2m + 5) \end{aligned}$$

If an expression looks like it might be the difference of two perfect squares, but the numbers don't work, look for a common factor. Always check for an HCF first.

Example

Factorise $2y^2 - 18$.

$$\begin{aligned} 2y^2 - 18 &= 2(y^2 - 9) \\ &= 2(y^2 - 3^2) \\ &= 2(y - 3)(y + 3) \end{aligned}$$

If an expression consists of one term subtracted from another, it is always worth checking whether it is the difference of two perfect squares. Even if the expressions look quite complicated (e.g. there might be a variable in the second term, or a power higher than 2), we can use the same rules.

Examples

Factorise these expressions.

a) $m^2 - 9n^2$

$$m^2 - 9n^2 = m^2 - (3n)^2$$
$$= (m - 3n)(m + 3n)$$

b) $4a^2 - 9b^2$

$$4a^2 - 9b^2 = (2a)^2 - (3b)^2$$
$$= (2a - 3b)(2a + 3b)$$

c) $(c + 2)^2 - 9$

$$(c + 2)^2 - 9 = (c + 2)^2 - 3^2$$
$$= ([c + 2] - 3)([c + 2] + 3)$$
$$= (c + 2 - 3)(c + 2 + 3)$$
$$= (c - 1)(c + 5)$$

d) $y^4 - 16$

$$y^4 - 16 = (y^2)^2 - 4^2$$
$$= (y^2 - 4)(y^2 + 4) \qquad (y^2 - 4) \text{ is also the difference}$$
$$= (y^2 - 2^2)(y^2 + 4) \qquad \text{of two perfect squares. So we}$$
$$= (y - 2)(y + 2)(y^2 + 4) \quad \text{can use the rule a second time.}$$

We can also use the rules with numerical calculations.

Examples

Use the rules of factorising to help you work out the answer to each of these calculations.

a) $(73)^2 - (27)^2$

$$(73)^2 - (27)^2 = (73 - 27)(73 + 27)$$
$$= 46 \times 100$$
$$= 4600$$

b) $(6.4)^2 - (3.6)^2$

$$(6.4)^2 - (3.6)^2 = (6.4 - 3.6)(6.4 + 3.6)$$
$$= 2.8 \times 10$$
$$= 28$$

Exercise 4

1 Factorise these expressions.

a) $a^2 - 49$ b) $b^2 - 25$ c) $16 - c^2$

d) $144 - d^2$ e) $e^2 - 81$ f) $f^2 - 36$

g) $100 - g^2$ h) $121 - b^2$ i) $4z^2 - 81$

j) $16y^2 - 25$ k) $49 - 100k^2$ l) $121 - 25x^2$

m) $25m^2 - 1$ n) $64n^2 - 1$ o) $5w^2 - 500$

p) $2p^2 - 32$ q) $108 - 3v^2$ r) $64 - 4r^2$

2 Factorise these expressions.

a) $a^2 - z^2$ b) $b^2 - y^2$ c) $c^2 - 4x^2$

d) $d^2 - 9w^2$ e) $16e^2 - v^2$ f) $25f^2 - u^2$

g) $16g^2 - 25t^2$ h) $49b^2 - 64s^2$ i) $(r + 1)^2 - 4$

j) $(x + 1)^2 - (y - 2)^2$ k) $9x^2 - (y - 3)^2$ l) $(a^2 - 4)^2 - 25$

m) $144 - (m + 3)^2$ n) $169 - (n - 4)^2$ o) $b^4 - y^4$

p) $d^4 - 81$ q) $r^4 - 1$ r) $4(2x - y)^2 - (x + 2y)^2$

3 Use the rules of factorising to help you work out the answer to each of these calculations.

a) $(88)^2 - (12)^2$ b) $(77)^2 - (23)^2$ c) $(51)^2 - (49)^2$

d) $(875)^2 - (125)^2$ e) $(9.8)^2 - (0.2)^2$ f) $(8.7)^2 - (1.3)^2$

g) $(6.54)^2 - (3.46)^2$ h) $(5.43)^2 - (4.57)^2$ i) $\left(\frac{3}{4}\right)^2 - \left(\frac{1}{4}\right)^2$

j) $\left(\frac{5}{8}\right)^2 - \left(\frac{3}{8}\right)^2$ k) $\dfrac{69^2 - 44^2}{4^2 + 3^2}$ l) $\dfrac{78^2 - 22^2}{8^2 + 6^2}$

Factorising quadratic expressions of the form $x^2 + bx + c$

A quadratic expression is any expression that has the form $ax^2 + bx + c$, where a, b and c are constants and $a \neq 0$. In other words, the variables in quadratic expressions have powers of 2 or smaller (no powers bigger than 2). We will look first at expressions where $a = 1$, so the general form will be $x^2 + bx + c$, where b and c are constants (they may be positive or negative).

If the expression is *not* a perfect square, and it is *not* the difference of perfect squares, we need to find another way to factorise it.

Let us look at the quadratic expression $x^2 + 8x + 15$.

We know that the factors of this expression will look like this: $(x + p)(x + q)$, where p and q are numbers (usually they are different numbers, unless the expression is a perfect square).

Let us expand these two brackets.

$$(x + p)(x + q) = x^2 + px + qx + pq$$
$$= x^2 + (p + q)x + pq$$

We can now compare this to our expression.

x^2	$+$	$8x$	$+$	15
$\downarrow$		$\downarrow$		$\downarrow$
x^2	$+$	$(p + q)x$	$+$	pq

This means that $p + q = +8$ and, **at the same time**, $p \times q = +15$.

So, to factorise $x^2 + 8x + 15$, we must find two numbers that will add up to $+8$ **and** that will multiply to give $+15$.

If $p \times q = 15$, then p and q must be a **pair of factors** of 15.
15 has two factor pairs: 1×15 and 3×5
We also know that p and q must add up to $+8$.
$1 + 15 \neq 8$ but $3 + 5 = 8$, so the two numbers must be $+3$ and $+5$.

This gives us the factors $(x + 3)$ and $(x + 5)$.
$x^2 + 8x + 15 = (x + 3)(x + 5)$

In this expression, both signs are positive. Sometimes, one or more must be negative. Use these rules to help you choose the correct sign for each factor.

- If the **last term** in the expression is **positive**, then **both factors** will have the **same sign**.
  - If the **middle** term is **positive**, then both **factors** will have a **+ sign**. $(x^2 + bx + c) = (x + ?)(x + ?)$
  - If the **middle** term is **negative**, then both **factors** will have a **− sign**. $(x^2 - bx + c) = (x - ?)(x - ?)$
- If the **last term** in the expression is **negative**, then **one factor** will have a **+ sign** and the **other factor** will have a **− sign**. $\left. \begin{array}{l} x^2 + bx - c \\ x^2 - bx - c \end{array} \right\} = (x + ?)(x - ?)$

Use these steps to help you factorise a quadratic expression in the form $x^2 + bx + c$:

1　Write all terms in decreasing order of the variable powers (e.g. x^2, then x, then numbers).
2　Remember to take out any HCF of the expression first.
3　Use the rules (on page 172) to decide what sign must be in each of the factors.
4　Write the brackets with these signs and a space for the numbers: $(x \quad + \text{ or } - \quad \ldots)(x \quad + \text{ or } - \quad \ldots)$.
5　Write down all the pairs of factors of the last term in the expression.
6　Choose one pair of these factors so that you get the correct middle term when you multiply out your two brackets (be careful with the signs).

Examples

Factorise these quadratic expressions.
a) $x^2 + 3x + 2$

Both factors will have a positive sign: $(x + \ldots)(x + \ldots)$
Factors of 2: 1×2
So the factors will be $(x + 1)$ and $(x + 2)$.

So $x^2 + 3x + 2 = (x + 1)(x + 2)$

b) $x^2 - 8x + 12$

Both factors will have a negative sign: $(x - \ldots)(x - \ldots)$
Factors of 12: 1×12, 2×6, 3×4
The only pair that will make $-8x$ is -2 and -6.
So the factors will be $(x - 2)$ and $(x - 6)$.

So $x^2 - 8x + 12 = (x - 2)(x - 6)$

c) $x^2 + 2x - 15$

One factor will have a positive sign and one will have a negative sign: $(x + \ldots)(x - \ldots)$
Factors of 15: 1×15, 3×5
The only pair that will make $+2x$ is $+5$ and -3.
So the factors will be $(x + 5)$ and $(x - 3)$.

So $x^2 + 2x - 15 = (x + 5)(x - 3)$

d) $x^2 - 2x - 8$

One factor will have a positive sign and one will have a negative sign: $(x + \dots)(x - \dots)$

Factors of 8: $1 \times 8, 2 \times 4$

The only pair that will make $-2x$ is $+2$ and -4.

So the factors will be $(x + 2)$ and $(x - 4)$.

So $x^2 - 2x - 8 = (x + 2)(x - 4)$

Exercise 5

1 Copy and complete these equations.

a) $x^2 + 6x + 5 = (x + 5)(x \dots)$ b) $x^2 + 9x + 14 = (x + 7)(x \dots)$
c) $x^2 + 6x + 8 = (x + 4)(x \dots)$ d) $x^2 + 9x + 18 = (x + 6)(x \dots)$
e) $x^2 - 6x + 5 = (x - 5)(x \dots)$ f) $x^2 - 7x + 10 = (x - 5)(x \dots)$
g) $x^2 - 7x + 12 = (x - 3)(x \dots)$ h) $x^2 + 3x - 4 = (x + 4)(x \dots)$
i) $x^2 + 5x - 14 = (x + 7)(x \dots)$ j) $x^2 - 4x - 5 = (x - 5)(x \dots)$

2 Factorise each of these quadratic expressions.

a) $x^2 + 3x + 2$ b) $x^2 + 8x + 7$ c) $x^2 + 8x + 15$
d) $x^2 + 8x + 12$ e) $x^2 + 12x + 11$ f) $x^2 + 9x + 20$
g) $x^2 + 10x + 24$ h) $x^2 + 13x + 36$ i) $x^2 - 6x + 9$
j) $x^2 - 6x + 8$ k) $x^2 - 11x + 10$ l) $x^2 - 16x + 15$
m) $x^2 - 8x + 15$ n) $x^2 - 10x + 16$ o) $x^2 - 12x + 20$
p) $x^2 - 11x + 24$ q) $x^2 - x - 6$ r) $x^2 - 5x - 6$
s) $x^2 + 2x - 24$ t) $x^2 + 5x - 24$ u) $x^2 - 2x - 15$
v) $x^2 + 3x - 18$ w) $x^2 - 3x - 40$ x) $x^2 - 4x - 12$

3 Factorise each of these quadratic expressions.

a) $x^2 + 11x + 30$ b) $x^2 + 2x - 8$ c) $x^2 - 4x - 21$
d) $x^2 + x - 20$ e) $x^2 + 7x + 12$ f) $x^2 - 2x + 1$
g) $x^2 - 9x + 14$ h) $x^2 - 7x + 6$ i) $x^2 + 11x + 18$
j) $x^2 + 11x + 24$

Factorising quadratic expressions of the form $ax^2 + bx + c$ (when $a > 1$)

Now let us look at quadratic expressions where the coefficient of x^2 is bigger than 1.

When the coefficient of x^2 is **equal to 1**, we have seen that we need to find the factors of the last term only to work out the factors of the whole expression.

When the coefficient of x^2 is **bigger than 1**, we need to find the right factors for this term as well as for the last term, before we can find the correct factors of the whole expression.

> Use these steps to help you factorise a quadratic expression in the form $ax^2 + bx + c$ (when $a > 1$):
>
> 1. Write all terms in decreasing order of the variable powers (e.g. x^2 then x then numbers).
> 2. Remember to take out any HCF of the expression first.
> 3. Use the rules (page 172) to decide what sign must be in each of the factors.
> 4. Write the brackets with these signs and a space for the numbers: $(…x + \text{ or } - …)(…x + \text{ or } - …)$.
> 5. Write down all the pairs of factors of the coefficient of the x^2 term.
> 6. Write down all the pairs of factors of the last term in the expression.
> 7. Choose one pair from each set of factors so that you get the correct middle term when you multiply out your two brackets (be careful with the signs).
> 8. Check your factors by multiplying out the two brackets to see if their product gives the same expression as you started with.

NOTE: Step 7 is a matter of trial and error. There is no quick way to choose the right pairs of factors – you must try combining pairs of factors from each set again and again until you find the right ones. The more you practise this, the easier it becomes to choose the right factors.

Examples

a) Factorise the expression $5x^2 + 13x + 6$.

There is no HCF bigger than 1.

Both factors will have a positive sign: $(…x + …)(…x + …)$
Factors of 5: 1×5
Factors of 6: 1×6, 2×3

Now we need to choose one pair from each set of factors. Here is one way to decide.

Write the two lists of factors in a vertical list:

Factors of 5 (x^2 term)	Factors of 6 (last term)
1	1
5	6
	2
	3

Now try multiplying factors across lists and adding the answers together – keep trying until you get the right answer for the middle term (13). Always work in a logical order, step by step. Here we have only one pair of factors in the first list, but it can be combined with each pair in the second list in two different ways, making four possible combinations in total.

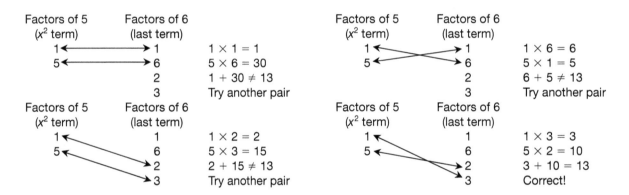

Now carefully position the numbers in the brackets so that $1x$ will multiply by 3, and $5x$ will multiply by 2 when you expand the brackets. A diagram like this one might help you:

$(1x + 2) \qquad (5x + 3)$

$+10x$

$+3x$

$+13x$

The sum of the inner and outer products must be equal to the middle term.

Finally, check your answer by multiplying out the brackets:
$(x + 2)(5x + 3) = 5x^2 + 13x + 6.$

So $5x^2 + 13x + 6 = (x + 2)(5x + 3)$

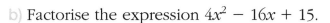

b) Factorise the expression $4x^2 - 16x + 15$.

There is no HCF bigger than 1.

Both factors will have a negative sign: $(...x - ...)(...x - ...)$
Now list the factors of 4 and 15 and try each combination in turn.

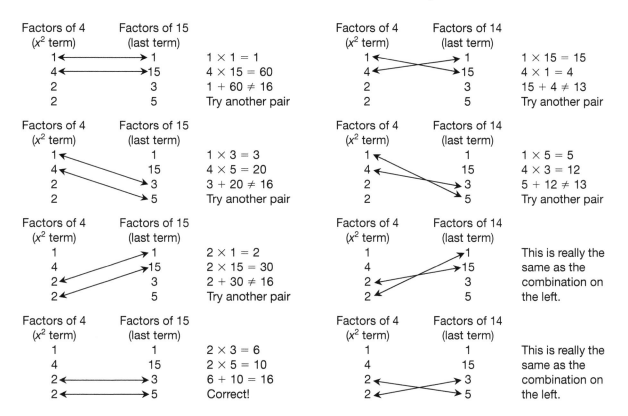

Now carefully position the numbers in the brackets so that one $2x$ will multiply by 3, and the other $2x$ will multiply by 5 when you expand the brackets.

$(2x - 5) \qquad (2x - 3)$

$-10x$

$-6x$

$-16x$

Finally, check your answer by multiplying out the brackets:
$(2x - 5)(2x - 3) = 4x^2 - 16x + 15$.

So $4x^2 - 16x + 15 = (2x - 5)(2x - 3)$

c) Factorise the expression $4x^2 + 19x - 5$.

There is no HCF bigger than 1.

One of the factors will have a positive sign and the other will have a negative sign: $(\ldots x + \ldots)(\ldots x - \ldots)$

Now list the factors of 4 and 5 and try each combination in turn. This time the brackets have opposite signs, so we must **subtract** one product from the other to make 19.

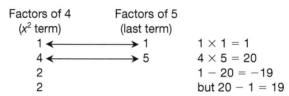

Factors of 4 (x^2 term)	Factors of 5 (last term)	
1 ⟷	1	$1 \times 1 = 1$
4 ⟷	5	$4 \times 5 = 20$
2		$1 - 20 = -19$
2		but $20 - 1 = 19$

The first combination is correct. To get $20 - 1 = 19$, one of the 1s must be negative.

Carefully position the numbers in the brackets so that $1x$ will multiply by -1, and $4x$ will multiply by 5 when you expand the brackets.

$$(4x - 1) \qquad (x + 5)$$
$$\begin{array}{c} \lfloor \quad -x \quad \rfloor \\ \lfloor\!\_\_\_\_\_ +20x \_\_\_\_\_\rfloor \\ \hline +19x \end{array}$$

Finally, check your answer by multiplying out the brackets:
$(4x - 1)(x + 5) = 4x^2 + 19x - 5$.

So $4x^2 + 19x - 5 = (4x - 1)(x + 5)$

d) Factorise the expression $-8x^2 + 14x + 15$.

When the x^2 term has a minus sign, it can make it more difficult to choose the correct factors with the correct signs. To avoid this, we can take out -1 as the HCF.

$-8x^2 + 14x + 15 = -1(8x^2 - 14x - 15)$.

We can now find the factors of the expression inside the brackets. At the end, we must remember to multiply our answer by -1.

One of the factors of $8x^2 - 14x - 15$ will have a positive sign and the other will have a negative sign: $(\ldots x + \ldots)(\ldots x - \ldots)$

Now list the factors of 8 and 15 and try each combination in turn. Remember to subtract rather than add.

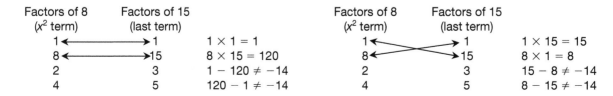

Factors of 8 (x^2 term)	Factors of 15 (last term)			Factors of 8 (x^2 term)	Factors of 15 (last term)	
1 ⟷	1	$1 \times 1 = 1$		1	1	$1 \times 15 = 15$
8 ⟷	15	$8 \times 15 = 120$		8	15	$8 \times 1 = 8$
2	3	$1 - 120 \neq -14$		2	3	$15 - 8 \neq -14$
4	5	$120 - 1 \neq -14$		4	5	$8 - 15 \neq -14$

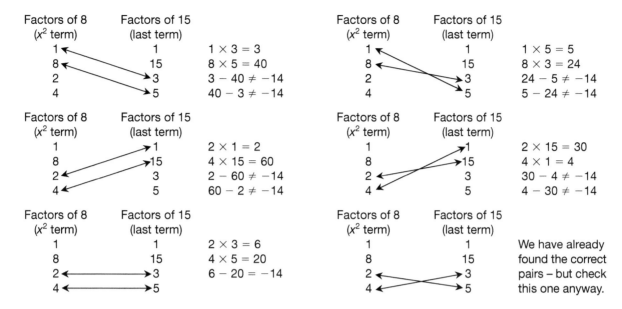

To get $6 - 20 = -14$, the 5 must be negative.

Carefully position the numbers in the brackets so that $2x$ will multiply by 3, and $4x$ will multiply by -5 when you expand the brackets.

$(2x - 5) \qquad (4x + 3)$

$$\underset{-14x}{\underline{\underset{+6x}{\underline{\quad -20x \quad}}}}$$

Now we must put the HCF of -1 back in.

$-1(8x^2 + 14x + 15) = -1(2x - 5)(4x + 3)$

We can write these factors in several different ways.

$-1(2x - 5)(4x + 3) = (-2x + 5)(4x + 3) = (5 - 2x)(4x + 3)$
$$= (2x - 5)(-4x - 3)$$

Check your answer by multiplying out each pair of factors:

$(-2x + 5)(4x + 3) = -8x^2 + 14x + 15$
$(5 - 2x)(4x + 3) = -8x^2 + 14x + 15$
$(2x - 5)(-4x - 3) = -8x^2 + 14x + 15$

So $-8x^2 + 14x + 15 = (-2x + 5)(4x + 3) = (5 - 2x)(4x + 3)$
$$= (2x - 5)(-4x - 3)$$

Exercise 6

1 Copy and complete these equations.

a) $2x^2 + 12x + 18 = 2(x + 3)(x \ldots)$ b) $2y^2 - 9y - 18 = (y - 6)(\ldots y \ldots)$
c) $2a^2 + 5a - 18 = (a - 2)(\ldots a \ldots)$ d) $6d^2 - 11d + 3 = (2d - 3)(\ldots d \ldots)$
e) $4x^2 - 13x + 3 = (x - 3)(\ldots x \ldots)$ f) $4m^2 + 4m - 3 = (2m - 1)(\ldots m \ldots)$

2 Factorise these quadratic expressions.

a) $2x^2 + 5x + 2$ b) $3x^2 + 2x - 5$ c) $4x^2 - 16x + 15$
d) $5x^2 - 12x - 9$ e) $11x^2 + 12x + 1$ f) $5x^2 + 12x + 4$
g) $4x^2 - 9x + 2$ h) $9x^2 - 36x + 32$ i) $12c^2 + 23c - 24$
j) $15x^2 - 22x + 8$ k) $-6x^2 - x + 35$ l) $11x^2 + 10x - 1$
m) $36m^2 - 46m + 14$ n) $9x^2 - 32x + 28$ o) $8b^2 - 50b + 63$
p) $-3t^2 + 17t + 28$ q) $-6n^2 + 10 - 28n$ r) $-65x^2 + 87x - 28$

Mixed practice at factorising quadratic expressions

So far the expressions you have factorised have been arranged in groups. Usually, however, different kinds of expressions will be all mixed up together in one exercise.

Use these questions to help you work out how to factorise any expression.

1 Is there a **highest common factor**? If so, take it out.
2 Is the expression a **difference of two squares**? If so, use that rule to factorise it.
3 Is the expression a **perfect square**? If so, use that rule to factorise it.
4 Will the expression factorise into **two brackets**? If so, follow the steps you learned above to factorise it. Remember to check whether the coefficient of the x^2 term is 1 or bigger than 1.

Above all, always work logically!

Exercise 7

1 Factorise these expressions as much as you can.

a) $3x^2 + 12x$ b) $t^2 - 16$ c) $x^2 + 4x + 3$
d) $y^3 - y$ e) $2d^2 + 18$ f) $p^2 - q^2$
g) $a^2 - 2a$ h) $x^2 - 2x + 1$ i) $2y^3 - 8y$
j) $a^2 - 6a + 9$ k) $3m^2 - 12$ l) $v^2 - v - 6$
m) $ax^2 - ay^2$ n) $8 - 2x^2$ o) $4 - 4k + k^2$
p) $18 - 2x^2$ q) $50 - 2x^2$ r) $12x^2 - 27y^2$
s) $6a^2 - 3a$ t) $x^2 - 15x + 56$

2 Factorise these expressions as much as you can.

a) $3a^2 - 2a - 8$ b) $3x^2 + 31x + 10$ c) $7x^2 - 19x + 10$
d) $6x^2 - 25x - 25$ e) $8x^2 + 2x - 15$ f) $-6x^2 - 3x + 18$

Unit 10 Pythagoras' theorem

Key vocabulary

converse	hypotenuse	theorem
corollary	perpendicular	three dimensions

A Who was Pythagoras and what is a theorem?

Pythagoras was a Greek mathematician who lived around the time of 540 BC.

In maths we have already learned many **rules** or **laws**. These tell us what is true, and we can use them to work out many other things. When we use a rule or a law to prove that something new is also true, we call this a theorem.

Pythagoras is best known for his theorem about right-angled triangles. He was not the first person to discover this theorem – the Babylonians discovered it much earlier. However, Pythagoras is recognised as the first person to have **proved** the theorem. The Chinese also knew this theorem at about the same time as Pythagoras.

B Pythagoras' theorem

Pythagoras' theorem relates to right-angled triangles. In a right-angled triangle one of the angles is 90°, and so two sides of the triangle are perpendicular to each other.

The **longest** side of a right-angled triangle is the side **opposite the right angle**. This side is called the hypotenuse.

Pythagoras' theorem describes a connection between the hypotenuse and the shorter sides of a right-angled triangle – or, more accurately, between the area of the squares on the sides.

Pythagoras' theorem

For any right-angled triangle, the area of the square on the hypotenuse is equal to the sum of the areas of the squares on the two shorter sides.

If a and b are the lengths of the two shorter sides of a right-angled triangle and c is the length of the hypotenuse, we can write Pythagoras' theorem as $c^2 = a^2 + b^2$.

NOTE: The lengths of all three sides must be measured in the same units.

Pythagoras' theorem means that, in the diagram, the area of the two **shaded** squares added together is equal to the area of the biggest, **unshaded** square. That is, $a^2 + b^2 = c^2$.

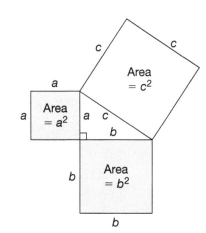

We can use Pythagoras' theorem in two ways.

- If we know the lengths of two sides of a right-angled triangle, we can use Pythagoras' theorem to find the length of the third side.
- If we know the length of each of the three sides of a triangle, we can use Pythagoras' theorem to prove that the triangle has a right-angle (or not). This is known as the converse of the theorem ('converse' means 'the opposite' or 'reverse').

C Finding the hypotenuse

To find the hypotenuse of a right-angled triangle, given the lengths of the two shorter sides, just substitute the given lengths into the theorem.

Example

PQR is a triangle with $\angle Q = 90°$, $PQ = 5\,\mathrm{cm}$ and $QR = 12\,\mathrm{cm}$.

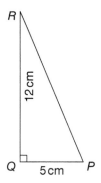

Calculate the length of *PR*.

$$PR^2 = PQ^2 + QR^2 \quad \text{(Pythagoras' theorem)}$$
$$PR^2 = 5^2 + 12^2$$
$$= 25 + 144$$
$$= 169$$
$$PR = \sqrt{169} = 13$$

PR is 13 cm long.

NOTE: We ignore the negative square root as length cannot be negative.

D Finding one of the shorter sides

To find one of the shorter sides of a right-angled triangle, we can re-arrange Pythagoras' theorem.

$$a^2 + b^2 = c^2$$

So $\quad a^2 = c^2 - b^2$

and $\quad b^2 = c^2 - a^2$

This is sometimes known as the corollary to Pythagoras' theorem.

(A 'corollary' is something that must be true because something else is already known to be true.)

So to find the length of either of the shorter sides of a right-angled triangle, subtract the square of the known shorter side from the square of the hypotenuse. The square root of the result will be the length of the remaining shorter side.

Examples

a) Calculate the length of side *a* in this right-angled triangle.

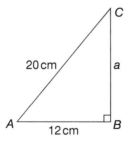

$$AB^2 + BC^2 = AC^2 \quad \text{(Pythagoras' theorem)}$$
$$12^2 + a^2 = 20^2$$
$$a^2 = 20^2 - 12^2$$
$$= 400 - 144 = 256$$
$$a = \sqrt{256} = 16$$

Side *a* is 16 cm long.

b) Calculate the length of side *r* in this diagram.

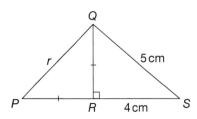

In the right-hand triangle:
$$QR^2 + RS^2 = QS^2$$
$$QR^2 + 4^2 = 5^2$$
$$QR^2 = 5^2 - 4^2$$
$$= 25 - 16$$
$$= 9$$
$$QR = \sqrt{9}$$
$$= 3$$
QR is 3 cm long.

In the left-hand triangle:
$$QR^2 + PR^2 = PQ^2$$
$$3^2 + 3^2 = r^2$$
$$r^2 = 18$$
$$r = \sqrt{18}$$
$$= 4.2 \text{ (to 1 d.p.)}$$
Side *r* is 4.2 cm long.

Exercise 1

1 Calculate the length of the hypotenuse in each of these right-angled triangles.

a)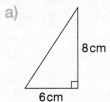

b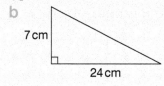

c)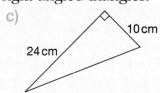

2 These triangles are right-angled. Calculate the length of side *a* correct to 1 decimal place.

a)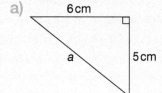

b)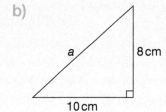

c)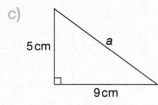

3 Work out the length of side *b* of each of these triangles.

a)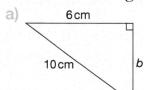

b)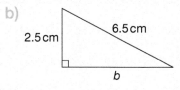

c)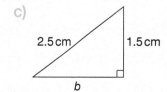

4 Work out the length of side *c* of each of these triangles, correct to 1 decimal place.

a)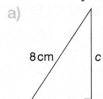

b)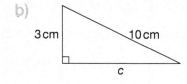

c)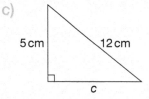

5 In the triangle shown, $\angle FGE = 90°$. For the following side lengths, calculate the length of the third side of the triangle.

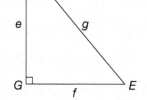

a) $e = 9\,\text{cm}, f = 12\,\text{cm}$ b) $f = 12\,\text{m}, g = 13\,\text{m}$

c) $e = 40\,\text{cm}, g = 41\,\text{cm}$ d) $e = 35\,\text{cm}, g = 37\,\text{cm}$

e) $e = 28\,\text{cm}, g = 53\,\text{cm}$ f) $e = 4\,\text{m}, f = 7\,\text{m}$

g) $g = 9\,\text{cm}, f = 6\,\text{cm}$ h) $g = 9.4\,\text{cm}, e = 4.6\,\text{cm}$

6 Look at this graph. Each unit on the *x*-axis and the *y*-axis is 1 cm.

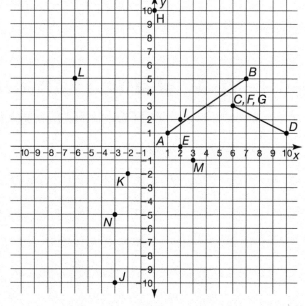

a) Use Pythagoras' theorem to calculate the exact length of the two lines *AB* and *CD*.

b) Using Pythagoras' theorem and the method you used in part **a)**, calculate the distance between each of these pairs of points on the graph.

  i) $E(2, 0)$ and $F(6, 3)$

  ii) $G(6, 3)$ and $H(0, 10)$

  iii) $I(2, 2)$ and $J(-3, -10)$

  iv) $K(-2, -2)$ and $L(-6, 5)$

  v) $M(3, -1)$ and $N(-3, -5)$

7 Calculate the length of side *a* in each of these diagrams.

a)
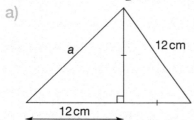

b)
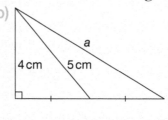

c)
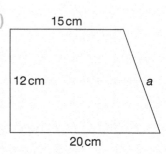

8 Calculate the length of sides *a* and *b* in each of these diagrams.

a)
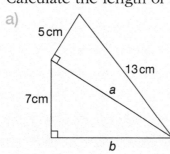

b)
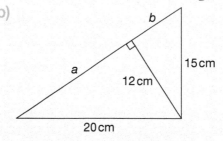

c)
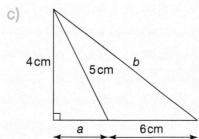

E Using Pythagoras' theorem to solve word problems

You need three things to solve a word problem successfully. You must:

- Understand the problem.
  What information is given?
  What are you asked to find?
- Draw diagrams.
  In some questions, a diagram is not given.
  Drawing a diagram will usually help you to see the problem more clearly.
- Choose a right-angled triangle that you can use.
  In more complicated problems, there may be many right-angled triangles, and you will have to choose the one that can be used to calculate the answer.

Sometimes you will have to make a right-angled triangle by dividing the shape into different parts with straight lines.
It is a good idea to draw the triangle you are going to use on its own – especially if it has been taken from part of a three-dimensional drawing.

Examples

a) The roof of a house is 12 m above the ground. To make it safe, the bottom of a ladder must be placed 5 m away from the wall.
How long must the ladder be to reach the roof safely?

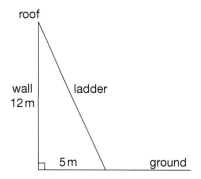

$$l^2 = 5^2 + 12^2 \quad \text{(Pythagoras' theorem)}$$
$$= 25 + 144$$
$$= 169$$
$$l = \sqrt{169}$$
$$= 13$$

The ladder must be 13 m long.

NOTE: We ignore the negative square root, as length cannot be negative.

b) The diagram shows a swimming pool 'cut' from side to side.
The bottom slopes evenly from 1 m deep to 4.2 m deep. The pool is 30 m long.
Find the length of the sloping bottom of the pool (correct to 3 significant figures).

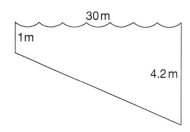

We can draw a line EC that is at right-angles to the side BD. This makes $\triangle CDE$ a right-angled triangle that we can use.
$$DE^2 = CD^2 + CE^2$$
$$= 3.2^2 + 30^2$$
$$= 10.24 + 900$$
$$= 910.24$$
$$DE = \sqrt{910.24}$$
$$= 30.170\,183\ldots$$

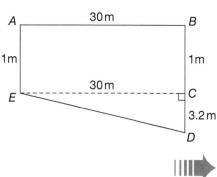

The length of the sloping bottom of the pool is 30.2 m (to 3 s.f.).

c) One diagonal of a rhombus is 12 cm long and the other diagonal is 16 cm long.
Calculate the length of the sides of the rhombus.
All sides of a rhombus are equal in length.
The diagonals, AC and BD intersect at their mid points E and at right angles.

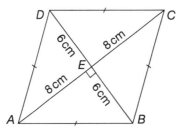

$AB^2 = AE^2 + BE^2$ (Pythagoras' theorem)
$AB^2 = 8^2 + 6^2$
$ = 64 + 36 = 100$
$AB = \sqrt{100} = 10$
The sides of the rhombus are each 10 cm long.

Exercise 2

1 The length of a rectangle is 24 cm. The diagonals of this rectangle are 26 cm long. Work out the area of the rectangle.

2 The diagonals of a square are 14 cm long.
Work out the area of this square.

3 One diagonal of a rhombus is 24 cm long. Each side of the rhombus is 13 cm long. Calculate the length of the second diagonal.

4 PQR is an equilateral triangle with $QR = 2$ cm.
Calculate the perpendicular distance from P to QR (to 1 decimal place).

5 The perpendicular height of an isosceles triangle is 12 cm. The base of the triangle is 18 cm long.
Calculate the length of the two equal sides.

6 $\triangle PQR$ has a right-angle at point Q. S is a point on QR.
$PS = 18$ cm, $QS = 9$ cm and $PR = \sqrt{468}$ cm.
Calculate the length of the side RS.

7 $ABCD$ is a rectangle with $AB = 10$ cm and $AD = 6$ cm.
E is a point on AB so that CDE is an isosceles triangle with $CD = CE$.
Calculate the length of DE (to 1 decimal place).

8 Calculate the area of the square *ABCD*.

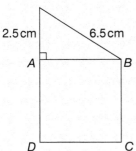

9 Two cars start from the same point and travel in different directions. One travels due East for 8 km and then stops. The other travels due South for 15 km and then stops.

How far apart are the two cars when they have stopped?

10 A rectangular field is 80 m long. The farmer wants to build a fence all around this field. The diagonal of the field is 100 m long.

a) Find the width of the field.

b) Find the length of the fence if it goes round the field completely.

11 An equilateral triangle has sides that are *x* cm long. The perpendicular height of this triangle is 12 cm long.

Calculate the perimeter of the triangle (to 1 decimal place).

12 The top of a lampshade has a diameter of 10 cm.

The bottom of the lampshade has a diameter of 20 cm.

The perpendicular height of the lampshade is 12 cm.

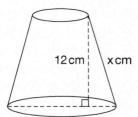

Calculate the length, *x*, of the sloping sides.

13 *ABCD* is a kite. *AB* = 8.5 cm, *BC* = 5.4 cm and *BD* = 7.6 cm.

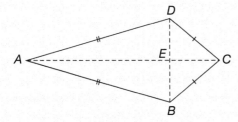

a) Calculate the length of *AC* (to 1 decimal place).

b) Calculate the area of the kite (to 1 decimal place).

F Solving problems in three dimensions

When solving problems in three dimensions, we sometimes need to use more than one triangle to help us find the length of the side we want.

It helps to draw each triangle we use separately. This way we can see just one part of the problem at a time.

Examples

a) This cube has sides of length 6 cm.

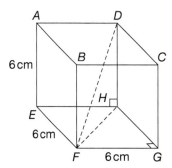

Calculate the distance from D to F (to 1 decimal place).

DF is the hypotenuse of the right-angled triangle $\triangle DFH$.
At the moment we only know the length of the side DH.
The length of the other side we need for the calculation, FH, can be found by using a different right-angled triangle $\triangle FGH$. FH is the hypotenuse of $\triangle FGH$.

$\triangle FGH$ looks like this:

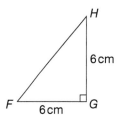

$$FH^2 = FG^2 + GH^2 \qquad \text{(Pythagoras' theorem)}$$
$$= 6^2 + 6^2$$
$$= 36 + 36$$
$$= 72$$
$$FH = \sqrt{72}$$

We don't need to calculate the value of $\sqrt{72}$, because in the next calculation we will be using the square of it.
We can now use the length FH in the next triangle to work out the side DH.

$\triangle DFH$ looks like this:

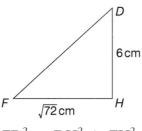

$FD^2 = DH^2 + FH^2$ (Pythagoras' theorem)
$FD^2 = 6^2 + (\sqrt{72})^2$
$\quad\;\; = 36 + 72$
$\quad\;\; = 108$
$FD = \sqrt{108}$
$\quad\;\; = 10.392304\ldots$

The length of the side FD is $10.4\,\text{cm}$ (1 d.p.).

b) $ABCDE$ is a pyramid with a rectangular base.

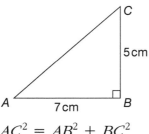

Calculate the perpendicular height, EF, of this pyramid, correct to 2 decimal places.

EF is one side of the right-angled triangle $\triangle CEF$, but we do not know the length of the side CF. We do know that CF is half of AC (diagonals of a rectangle bisect each other), and AC is the hypotenuse of the right-angled triangle $\triangle ABC$.

$\triangle ABC$ looks like this:

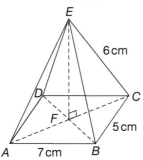

$AC^2 = AB^2 + BC^2$ (Pythagoras' theorem)
$AC^2 = 7^2 + 5^2$
$\quad\;\; = 49 + 25$
$\quad\;\; = 74$
$AC = \sqrt{74}\;\text{cm}$

So $CF = \frac{\sqrt{74}}{2}$ cm

Again, we don't need to calculate this value because in the next calculation, we will be using the square of it.

We can now use the length CF in the next triangle to work out the side EF.

$\triangle CEF$ looks like this:

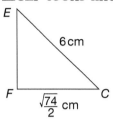

$EC^2 = CF^2 + EF^2$ (Pythagoras' theorem)

$6^2 = \left(\frac{\sqrt{74}}{2}\right)^2 + EF^2$

$EF^2 = 36 - \frac{74}{4}$

$= 36 - 18.5$

$= 17.5$

$EF = \sqrt{17.5}$

$= 4.183\,300\ldots$

The perpendicular height of the pyramid is 4.18 cm (2 decimal places).

 Exercise 3

In this exercise, calculate all answers correct to 3 significant figures.

1 A cone has a perpendicular height of 6 cm and a slant height of 6.5 cm.

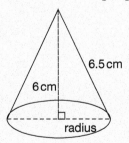

Calculate the radius of the circular base.

2 This cube has sides that are 7 cm long.

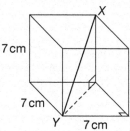

Calculate the length of the diagonal *XY*.

3 The wedge shape *PQRSTU* is shown in the diagram. *PS* = 15 cm, *ST* = 13 cm and *TR* = 6 cm.

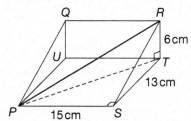

Calculate the length of the diagonal *PR*.

4 *ABCDE* is a pyramid with a rectangular base. *AB* = 6 cm, *BC* = 5 cm and the perpendicular height *EF* = 7 cm.

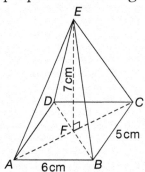

a) Calculate the length of *CF*.
b) Calculate the slant height *CE*.

5 *ABCDEFGH* is a rectangular cuboid.
AD = 10 cm, *CD* = 8 cm and *CG* = 6 cm.

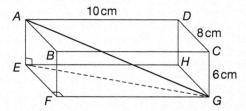

Calculate the length of the diagonal *AG*.

G Using the converse of Pythagoras' theorem

We have defined Pythagoras' theorem for any **right-angled** triangle. This means that we can use the Pythagorean relationship between sides of a triangle to **prove** whether the triangle is right-angled or not.

This is known as the **converse** of Pythagoras' theorem:

> If a, b and c are the lengths of the sides of any triangle (where c is the longest side), and $c^2 = a^2 + b^2$, then the triangle is a right-angled triangle.

Example

The sides of a triangle are given as $a = 12\,cm$, $b = 9\,cm$ and $c = 15\,cm$. Use the converse of Pythagoras' theorem to decide whether this triangle is right-angled or not.

$$a^2 + b^2 = 12^2 + 9^2$$
$$= 144 + 81$$
$$= 225$$
$$c^2 = 15^2$$
$$= 225$$
$$c^2 = a^2 + b^2, \text{ so the triangle is right-angled.}$$

Activity

Copy and complete this table. Use it to decide whether each of the triangles is right-angled or not.

	a	b	c	$a^2 + b^2$	c^2	$a^2 + b^2 = c^2$? Yes	$a^2 + b^2 = c^2$? No	$\triangle ABC$ right-angled? Yes	$\triangle ABC$ right-angled? No
a)	6	8	10	36 + 64	100	✓		✓	
b)	6	12	13						
c)	7	13	14						
d)	6	6.25	7.25						
e)	1.4	3.6	4						
f)	$2\frac{1}{2}$	6	$6\frac{1}{2}$						
g)	4	6.5	8.5						

Examples

a) $\triangle ABC$ has sides measuring 21 cm, 72 cm and 75 cm.
Use Pythagoras' theorem to decide whether this triangle has a right-angle.

Let c be the longest side, so $c = 75$.
The shorter sides are then a and b. Let us choose $a = 21$ and $b = 72$.

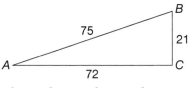

$$a^2 + b^2 = 21^2 + 72^2$$
$$= 441 + 5184$$
$$= 5625$$
$$c^2 = 75^2$$
$$= 5625$$

So $c^2 = a^2 + b^2$ and we can say that $\triangle ABC$ is a right-angled triangle.

b) $\triangle ACD$ and $\triangle BCD$ are two right-angled triangles that share the same side CD so that ADB is a straight line.
$AD = 9$ cm, $BD = 16$ cm and $CD = 12$ cm.
Prove that $\triangle ABC$ is also a right-angled triangle.

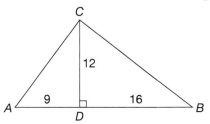

$\triangle BCD$ is a right-angled triangle, so using Pythagoras' theorem:
$$BC^2 = CD^2 + BD^2$$
$$= 12^2 + 16^2$$
$$= 144 + 256$$
$$= 400$$

$\triangle ACD$ is also a right-angled triangle, so using Pythagoras' theorem:
$$AC^2 = CD^2 + AD^2$$
$$= 12^2 + 9^2$$
$$= 144 + 81$$
$$= 225$$

In $\triangle ABC$, $AC^2 + BC^2 = 225 + 400$
$$= 625$$

For side AB we know that:
$$AB = AD + BD$$
$$= 9 + 16$$
$$= 25$$

So $AB^2 = 25^2$
$= 625$
So $AB^2 = AC^2 + BC^2$
This shows that $\triangle ABC$ is a right-angled triangle with $\angle ACB = 90°$.

c) $\triangle PQM$ and $\triangle PRM$ are two right-angled triangles that share the same side PM so that QMR is a straight line.
$PM = 8$ cm, $PQ = 17$ cm and $MR = 6$ cm.
Prove whether $\triangle PQR$ is a right-angled triangle or not.

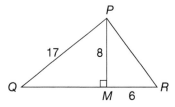

$\triangle PMR$ is a right-angled triangle, so using Pythagoras' theorem:
$PR^2 = PM^2 + MR^2$
$= 8^2 + 6^2$
$= 64 + 36$
$= 100$
$\triangle PQM$ is also a right-angled triangle, so using Pythagoras' theorem:
$PQ^2 = PM^2 + QM^2$
$17^2 = 8^2 + QM^2$
$QM^2 = 17^2 - 8^2$
$= 289 - 64$
$= 225$
$QM = 15$
For side QR we know that:
$QR = QM + MR$
$= 15 + 6$
$= 21$
So $QR^2 = 21^2$
$= 441$
In $\triangle PQR$, $PR^2 + PQ^2 = 100 + 289$
$= 389$
So $QR^2 \neq PR^2 + PQ^2$
This shows that $\triangle PQR$ is not a right-angled triangle.

Exercise 4

1 The lengths of the sides of some triangles are given below.
For each triangle, use Pythagoras' theorem to decide whether the triangle is right-angled or not.

a) 6, 8, 10   b) 4, 6, 8   c) 8, 10, 12

d) 8, 17, 15   e) 1, 4, 6   f) 0.3, 0.4, 0.5

2 In each of these cases, use Pythagoras' theorem to prove whether $\triangle ABC$ is a right-angled triangle.

a) 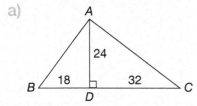   b)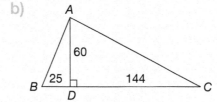

3 In triangle ABC the line CD is perpendicular to the line AB and meets it at the point D.
The lengths of some of the sides are given below.
Use these lengths and Pythagoras' theorem to prove whether $\triangle ABC$ is right-angled in each case.

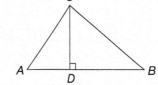

a) $AC = 13\,\text{cm}$, $BC = 15\,\text{cm}$ and $CD = 12\,\text{cm}$

b) $AC = 3\,\text{cm}$, $BC = 4\,\text{cm}$ and $CD = 2.4\,\text{cm}$

4 In the diagram, $AB = 21\,\text{cm}$, $BC = 28\,\text{cm}$, $CD = 7.2\,\text{cm}$, $DE = 9.6\,\text{cm}$ and $AE = 37\,\text{cm}$.
Use these lengths and Pythagoras' theorem to prove whether $\triangle ACE$ is a right-angled triangle.

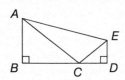

5 Calculate the length of the line AD (to 2 decimal places).

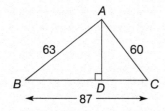

HINT: Let $DC = x$.

Unit 11

 Parallel lines, triangles and polygons

Key vocabulary

alternate angles	hexagon	pentagon
congruent	interior angles	polygon
corresponding angles	intersect	regular polygon
corresponding sides	irregular polygon	supplementary
decagon	nonagon	transversal
exterior angle	octagon	vertically opposite
heptagon	parallel	

 ## A Some angle facts

Revision

We have already learned that:

- the sum of all angles on a straight line is 180° (they are supplementary)

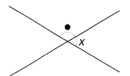

- when two straight lines cross, vertically opposite angles are equal.

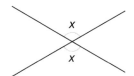

 ## B Parallel lines

Lines that run side by side, without ever getting any nearer or further apart from each other, are called parallel lines. Parallel lines can never intersect (or cross).

There are many different examples of parallel lines all around us:

- the sides of your ruler
- railway lines
- the sides of doors or windows (usually!)
- the opposite edges of a TV screen etc....

To show that lines are parallel in a diagram, we draw matching pairs of arrows on the lines that are parallel.

A straight line that cuts through parallel lines is called a transversal.

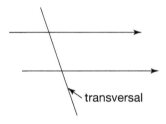

transversal

The angles made by two parallel lines and a transversal

Activity

The diagram shows a pair of parallel lines with a transversal cutting them. Some of the angles formed by these lines have been labelled.

Use your protractor to measure the sizes of each of the marked angles. Copy and complete the table.

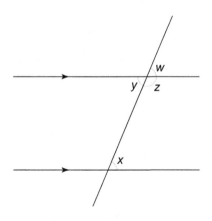

Angle	w	x	y	z
Degrees				

Corresponding angles

The two angles labelled w and x are called corresponding angles.

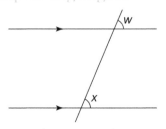

Four angles are formed at the point where the transversal crosses each of the parallel lines.

Each of the four angles formed around the point where the transversal crosses the **first parallel line** have a **corresponding angle** at the same place around the point where the transversal crosses the **second parallel line**.

Copy the diagram and mark each of the other pairs of corresponding angles. Use a different symbol for each of the pairs.

Now look at the table you completed earlier. You will see that the size of angle x is the same as the size of angle w.

> Pairs of **corresponding angles** are equal in size.

Angles between parallel lines

Four angles lie between the parallel lines. Two of these angles are on one side of the transversal, and the other two angles are on the other side of the transversal.

The two angles marked x and y are called alternate angles.

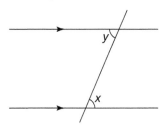

Pairs of alternate angles lie **between** the two **parallel lines**, but on **opposite sides of the transversal**.

Copy the diagram and mark the other pair of alternate angles, using a different symbol.

Look at the table you completed earlier. You will see that the size of angle x is the same as the size of angle y.

> Pairs of **alternate angles** are equal in size.

The two angles marked x and z are called interior angles.

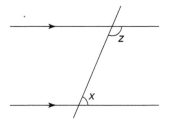

Pairs of interior angles lie **between** the two **parallel lines**, and on **the same side of the transversal**.

Copy the diagram and mark the other pair of interior angles, using a different symbol.

Look at the table you completed earlier. You will see that the size of angle x plus the size of angle z is 180° – they are **supplementary angles**.

> Pairs of interior angles are **supplementary** (add up to 180°).

These three rules are true for **any transversal** cutting **any two parallel lines**. This means that we can use these rules to **prove** whether two lines are parallel: if any one of the rules is true, the lines are parallel.

If a transversal cuts two lines, then the two lines are **parallel** if:

- a pair of **corresponding** angles is equal; *or*
- a pair of **alternate** angles is equal; *or*
- a pair of **interior** angles is supplementary (adds up to 180°).

Examples

a) Look at the angles marked in this diagram.

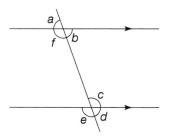

Name all the pairs of:
i) corresponding angles

$\angle b$ and $\angle d$
$\angle e$ and $\angle f$

ii) alternate angles

$\angle f$ and $\angle c$

iii) interior angles.

$\angle b$ and $\angle c$

b) Find the size of $\angle a$, $\angle b$ and $\angle c$ in each of the diagrams.

i)

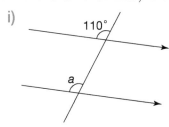

$\angle a = 110°$ (corr. $\angle$s, // lines)

ii)

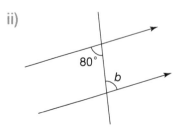

∠b = 80° (alt. ∠s, // lines)

iii)

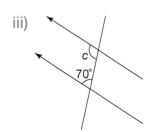

∠c = 180° − 70°
     = 110° (int. ∠s, // lines)

NOTE: In geometry, it is important that we always tell everyone which **rule** we have used to work out each answer. To do this, we write an abbreviation of the rule in brackets after each answer.

Using the three rules we have already learned, we can prove two more rules.

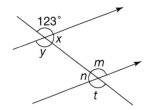

In the diagram, ∠m = 123° (corr. ∠s, // lines)
Also, ∠x = 180° − ∠m (int. ∠s, // lines)
            = 180° − 123°
            = 57°
So ∠n = 57° (alt. ∠s, // lines)

Also, ∠y = 180° − ∠n (int. ∠s, // lines)
            = 180° − 57°
            = 123°
So ∠t = 123° (corr. ∠s, // lines)

Now we can see that ∠m = 123° = ∠t.
This proves that **vertically opposite angles are equal**.

We can also see that ∠t + ∠n = 180°.
This proves that the **sum of angles on a straight line is 180°**.

Examples

a) Find the size of $\angle s$, $\angle v$ and $\angle w$ shown in the diagram.

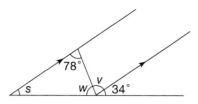

$\angle s = 34°$ (corr. $\angle$s, // lines)
$\angle v = 78°$ (alt. $\angle$s, // lines)
$\angle w = 180° - (34° + \angle v)$ ($\angle$s on a straight line)
$\quad = 180° - (34° + 78°)$
$\quad = 180° - 112°$
$\quad = 68°$

b) Two straight lines, AB and ED are parallel.

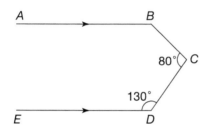

If $\angle EDC = 130°$ and $\angle DCB = 80°$, calculate $\angle ABC$.

The first thing to do is to draw another line, FC, so that FC is parallel to AB and ED.
This will divide $\angle BCD$ into two angles that we can call x and y.

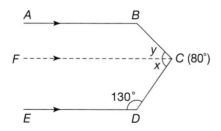

Note that line BC is now a transversal cutting the two parallel lines AB and FC.
Also, line CD is a transversal cutting the two parallel lines FC and ED.
So now we can use our rules for parallel lines:
$\quad \angle x = 180° - 130° = 50°$ (int. $\angle$s, $FC \parallel ED$)
$\angle x + \angle y = 80°$,
$\quad$ so $\quad \angle y = 80° - 50° = 30°$
$\quad \angle ABC = 180° - 30°$ (int. $\angle$s, $AB \parallel FC$)
$\quad\quad\quad = 150°$

c) The lines *QP* and *TS* are parallel.

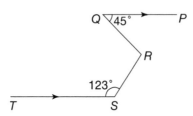

If $\angle PQR = 45°$ and $\angle RST = 123°$, calculate $\angle QRS$.

The first thing to do is to draw another line, *UR*, so that *UR* is parallel to *QP* and *TS*.
This will divide $\angle QRS$ into two angles that we can call *a* and *b*.

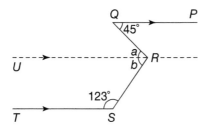

Note that line *QR* is now a transversal cutting the two parallel lines *QP* and *UR*.
Also, line *RS* is a transversal cutting the two parallel lines *UR* and *TS*.
So now we can use our rules for parallel lines:

$\angle a = 45°$ (alt. $\angle$s, $UR \parallel QP$)
$\angle b = 180° - 123°$ (int. $\angle$s, $UR \parallel TS$)
$\quad = 57°$

$\angle QRS = \angle a + \angle b$
$\quad\quad\quad = 45° + 57°$
$\quad\quad\quad = 102°$

Exercise 1

1 Write down the relationship between the pairs of angles given below.

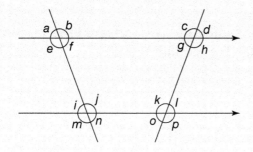

Give a reason for each answer.

For example: *a* and *i* are equal because they are corresponding angles.

a) *b* and *e*   b) *h* and *p*   c) *b* and *j*   d) *f* and *i*
e) *i* and *j*   f) *f* and *j*   g) *h* and *k*   h) *o* and *l*
i) *d* and *h*   j) *g* and *k*   k) *g* and *o*   l) *e* and *j*

2 Calculate all the angles marked with letters. Give reasons for your answers.

a)

b)
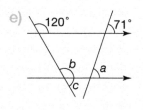

c)
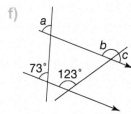

d)
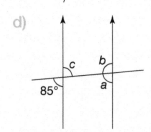

e)
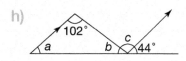

f)
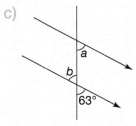

g)
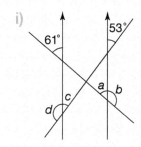

h)
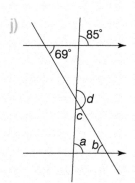

i)
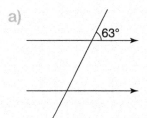

j)
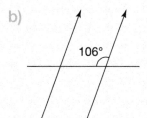

3 Copy the diagrams and write down the sizes of all the angles you can work out.

a)
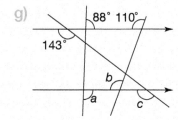

b)

c)
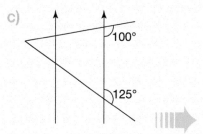

d)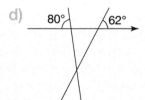

e)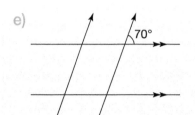

f)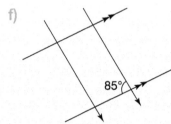

4 Calculate the angle marked with a letter in each diagram. Give reasons for your answers.

a)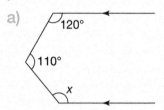

b)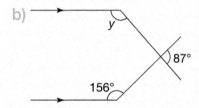

c)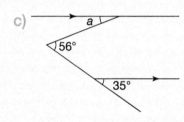

d)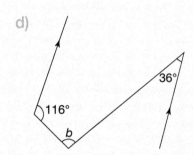

5 Calculate the angles marked with letters in each diagram. Give reasons for your answers.

a)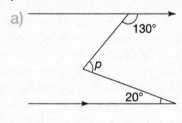

b)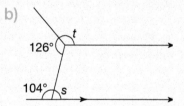

c)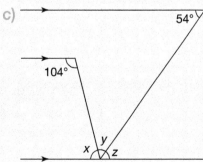

6 A farm has some fields with parallel sides. Using the angles given, calculate all the other angles in the fields. Copy the diagram and write the size of each angle on it.

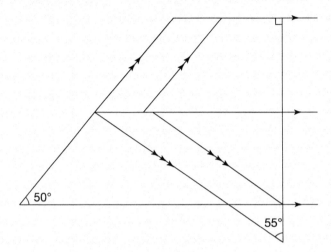

50°

55°

7 In the diagram, you are given the sizes of three angles: ∠QPF = 122°, ∠CAB = 26° and ∠ABC = 32°.

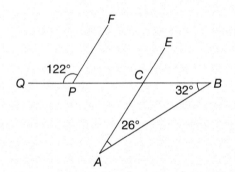

Prove that the line PF is parallel to the line AE.

Constructing parallel lines

We can construct a line PQ that is parallel to a given line ST.

Follow these steps:

1 Draw the first line ST (about 15 cm long will be OK).
2 Open a pair of compasses so that the points are about 4–5 cm apart.
3 Place the compass point on the line you have drawn, at any point near one end.
4 Draw an arc with the compass above the line.

5 Move the compass point about 3 cm along the line, and draw another arc to intersect with the first one.

6 Move the compass point about 2 cm along the line and draw another arc above the line.

7 Move the compass point another 3 cm along the line, and draw an arc to intersect with the one drawn in step 6 above.

8 Draw another pair of intersecting arcs in the same way as described in steps 6 and 7.

9 Use a ruler to draw a line joining the three points where the arcs intersect. This line (we call it *PQ*) is parallel to the line *ST*.

C Triangles

The sum of angles inside a triangle

We already know that the sum of all three angles inside a triangle is 180°. Let us now **prove** this fact by using only our knowledge of parallel lines.

Consider any triangle *ABC* with the side *AC* extended on both sides past the ends of the triangle.

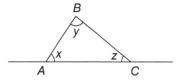

Now construct line *ED* through point *B* and parallel to the line *AC*.

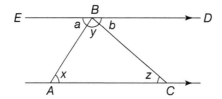

We know that $\angle a + \angle y + \angle b = 180°$ because they are on a single straight line.

We also know that $\angle a = \angle x$ because they are alternate angles between the parallel lines AC and ED with AB as transversal. In the same way, $\angle b = \angle z$ because they are alternate angles between the parallel lines AC and ED with BC as transversal.

Therefore we can say that $\angle x + \angle y + \angle z = 180°$.

> **Triangle rule 1**
> The interior angles of a triangle add up to $180°$.

The exterior angle of a triangle

If we extend any side of a triangle, the angle made outside the triangle is called the exterior angle of the triangle.

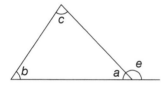

Here, $\angle e$ is the exterior angle.

We know that $\angle a = 180° - \angle e$ (sum of angles on a straight line). We also know that $\angle a = 180° - (\angle b + \angle c)$ (sum of angles in a triangle).

So $\angle e = \angle b + \angle c$

> **Triangle rule 2**
> The exterior angle equals the sum of the two opposite interior angles.

Examples

Find the value of x in each of these figures.

a)

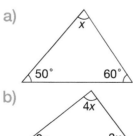

$$x = 180° - (50° + 60°)$$
$$= 180° - 110°$$
$$= 70°$$

b)

$$2x + 3x + 4x = 180°$$
$$9x = 180°$$
$$\frac{9x}{9} = \frac{180°}{9}$$
$$x = 20°$$

c)

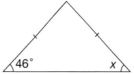

This is an isosceles triangle.
So $x = 46°$ (the base angles of an isosceles triangle are equal).

d)

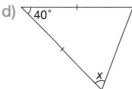

This is an isosceles triangle.
So $x = \dfrac{180° - 40°}{2}$ (Sum of angles in an isosceles triangle)

$= 70°$

e)

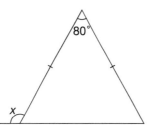

Let the base angles of the isosceles triangle be y (they are equal).

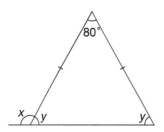

$y + y = 180° - 80°$ (Angles in a triangle add up to 180°.)
$\quad\;\; 2y = 100°$
$\quad\;\;\; y = 50°$

So $x = 180° - 50°$ (Adjacent angles on a straight line)
$\quad\;\; = 130°$

Exercise 2

1 Find the value of x in each of these figures.

a)
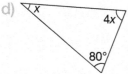
60°
2x   2x

b)
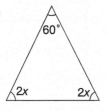
3x
3x   3x

c)
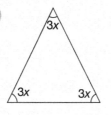
20°
4x   4x

d)

x
4x
80°

e)
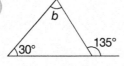
3x
x   2x

f)
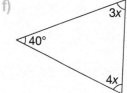
3x
40°
4x

2 ABC is an isosceles triangle that has the sides $\overline{AB} = \overline{AC}$ and $\angle BAC = 42°$. The line AD is parallel to the line BC. The line CAE is a straight line.

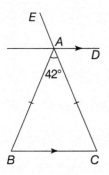

E
A   D
42°
B   C

Calculate the size of these angles.

a) $\angle BCA$ b) $\angle EAD$

3 Calculate the size of the unknown angle in each of these diagrams.

a)
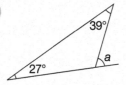
39°
27°   a

b)
39°...
b
30°   135°

c)
74°
112°   c

4 The straight lines BP and ACQ are parallel.

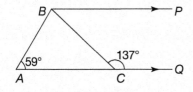

B   P
59°   137°
A   C   Q

a) Calculate $\angle CBP$.   b) Calculate $\angle ABC$.

5 The lines *AB* and *CDEF* are parallel.

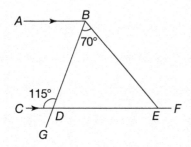

a) Calculate ∠*GDE*.   b) Calculate ∠*ABD*.   c) Calculate ∠*BEF*.

6 The line *XY* is parallel to the line *PQ*.
△*ABE* is an isosceles triangle.

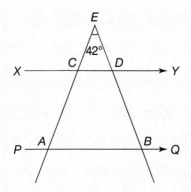

a) Name an angle that is equal to ∠*EAB*.
b) If ∠*CED* = 42°, calculate the size of the angle you have named in
   a) above.

7 In the diagram, the line *EF* is parallel to the line *BC* and ∠*BAE* = ∠*CAF*.

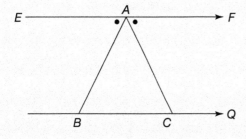

Prove that △*ABC* is an isosceles triangle.

Congruent triangles

Look at the picture of these two people. What do you notice about them?

Yes, they both look exactly the same! We call two people who look exactly the same **twins** – or, more accurately, **identical twins** (there are also twins that do not look the same).

In maths, we often find figures (or shapes) that are exactly the same size and shape. We can say that they are identical in every way, but we do **not** call them 'twins' – this word is used only for identical **people**! The word we use in maths for shapes that are identical in every way is congruent.

For example, if we have two triangles that are exactly the same size and shape, we call them **congruent triangles**. The size of the three angles in one of these triangles will be exactly the same size as the three angles in the other triangle. This is also true for the three sides of both triangles.

We can use these facts to help us work out the size of angles and lengths of sides in triangles that are congruent. Or, if we know the size of the angles and the lengths of the sides, we can say which triangles are congruent and which ones are not congruent.

Example

Are these triangles congruent?

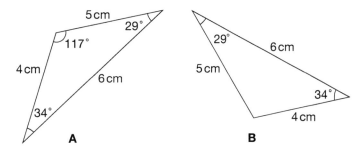

We can see that the missing angle in triangle **B** is 117°. (Two of the angles in triangle **A** are the same as the two angles we are given in triangle **B**. Since the angles in a triangle must add up to 180°, the third angle must be the same in both triangles.)
The three angles in triangle **A** are exactly equal to the three angles in triangle **B**.

The three sides of triangle **A** are also exactly the same length as the three sides of triangle **B**.

So we can say that triangle **A** and triangle **B** are **congruent**.

The short way to write this in maths is: $\triangle A \equiv \triangle B$. The symbol '$\equiv$' means 'is congruent to'.

Sometimes, triangles look congruent, but we cannot be sure because we do not know the size of every angle or the length of every side. This means that we need some rules that will help us to decide if triangles are congruent or not.

Rule 1

If we know the length of each of the three sides, the triangle we draw with these three sides will always be exactly the same size and shape. (Check that this is true: draw a triangle with sides 3 cm, 5 cm and 7 cm. Draw it as many times as you can to see if any of the triangles are any different in size or shape.)

This means that if we know the three sides of one triangle are exactly the same lengths as the three sides of another triangle, then these two triangles are **congruent**.

> **Congruent rule 1 – the Side-Side-Side rule (SSS)**
> Two triangles are congruent if **all three sides** of one triangle are exactly the same length as **all three sides** of the other triangle.

Rule 2

Let's construct a triangle ABC so that $\overline{AB} = 4\,\text{cm}$, $\angle A = 80°$ and $\angle B = 30°$.

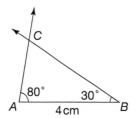

When we measure and draw angle A and angle B we find that they always cross at the same point, C. Point C will always be in the same place, and this means that the line segments $\overline{AC}$ and $\overline{BC}$ will always be the same length, and $\angle C$ will always be the same size. (Check for yourself that this is true.)

This means that every triangle we draw with one side 4 cm long, and angles of 80° and 30° **at each end of this side**, will be exactly the same size and shape, or congruent.

This is true whatever the length of the side, and whatever the size of the angles at each end (although to make a triangle, of course, at least one of them must be an acute angle).

> **Congruent rule 2 – the Angle-Side-Angle rule (ASA)**
> If two angles and the side **between** these angles in one triangle are equal in size to two angles and the side **between** them in a second triangle, then these two triangles are congruent.

Rule 3

Now let's construct a triangle with $\angle A = 80°$, side $\overline{AB} = 3$ cm and side $\overline{AC} = 2$ cm.

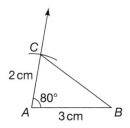

When we start with $\overline{AB}$, then draw $\angle A$ and $\overline{AC}$, we find that point C will always be in the same place. This means that the length of $\overline{BC}$ and the size of $\angle B$ and $\angle C$ will also always be the same. (Check for yourself that this is true.)

This means every triangle we draw with one side 3 cm long, another side 2 cm long and the angle **between these two sides** equal to 80° will be exactly the same size and shape, or congruent.

This is true whatever the length of two sides and whatever the angle between them (although to make a triangle, of course, it must be less than 180°).

> **Congruent rule 3 – the Side-Angle-Side rule (SAS)**
> If two sides and the angle **between** them in one triangle are equal to two sides and angle **between** them in a second triangle, then these two triangles are congruent.

Rule 4

We already know that if two angles and the side between these angles are the same size in two triangles, then these two triangles are congruent (ASA).

We also know that if any two angles in a triangle are the same size as two angles in another triangle, then the third angle in both triangles must also be equal.

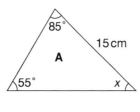

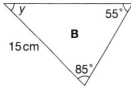

In triangle **A**
$\angle x = 180° - (85° + 55°)$ (Angles in a triangle add up to 180°.)
    $= 40°$

In triangle **B**
$\angle y = 180° - (85° + 55°)$ (Angles in a triangle add up to 180°.)
    $= 40°$

So $\angle x = \angle y = 40°$

If we also know that at least one pair of corresponding sides is the same length, then we can use the ASA rule to prove that triangle **A** is congruent to triangle **B**.

NOTE: **Corresponding sides** in two triangles join angles that are the same size in the two triangles.

This is true for any two triangles with two angles equal and the corresponding side equal.

> **Congruent rule 4 – the Angle-Angle-Side rule (AAS)**
> If two angles in a triangle are equal to two angles in another triangle, and at least one pair of corresponding sides is the same length, then the two triangles are congruent.

NOTE: Any two triangles that are congruent because of an Angle-Side-Angle relationship (congruent rule 2), can also be proved congruent using the Angle-Angle-Side relationship (congruent rule 4).

Isosceles triangles

$\triangle ABC$ is an **isosceles** triangle.

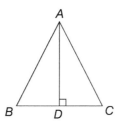

This means that $\overline{AB} = \overline{AC}$ and $\angle ABD = \angle ACD$.
The line $\overline{AD}$ is drawn so that $\overline{AD}$ is perpendicular to $\overline{BC}$.
So $\angle ADB = \angle ADC = 90°$.
So $\triangle ABD \equiv \triangle ACD$ (AAS)
So $\overline{BD} = \overline{CD}$ and $\angle BAD = \angle CAD$

> **Triangle rule 3**
> For any isosceles triangle, a line drawn from the vertex and perpendicular to the base, bisects (cuts in half) the base and also the angle at the vertex.

Examples

a) Which two of these triangles are congruent? Give a reason for your answer.

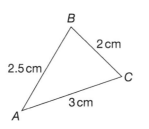
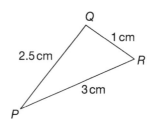
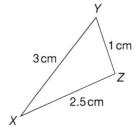

$PQ = XZ$, $QR = ZY$ and $RP = YX$
So $\triangle PQR \equiv \triangle XYZ$ (SSS)

b) Which two of these triangles are congruent? Give a reason for your answer.

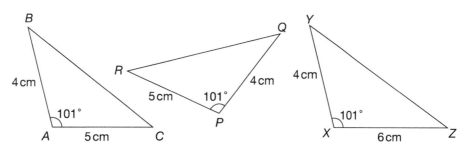

$AB = PQ$, $\angle A = \angle P$ and $AC = PR$
So $\triangle ABC \equiv \triangle PQR$ (SAS)

c) Which two of these triangles are congruent? Give reasons for your answer.

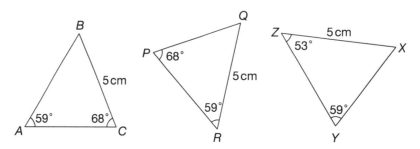

Solution 1

$\angle B = 180° - (59° + 68°)$ (Angles in a triangle add up to 180°.)
  $= 53°$

$\angle X = 180° - (59° + 53°)$ (Angles in a triangle add up to 180°.)
  $= 68°$

So $\angle B = \angle Z$, $BC = XZ$ and $\angle C = \angle X$

So $\triangle ABC \equiv \triangle XYZ$ (ASA)

Solution 2

$\angle B = 180° - (59° + 68°)$ (Angles in a triangle add up to 180°.)
  $= 53°$
  $= \angle Z$

$BC = XZ$ (These are **corresponding sides** as they are both
    opposite the angle of 59°.)

So $\angle B = \angle Z$, $BC = XZ$ and $\angle A = \angle Y$

So $\triangle ABC \equiv \triangle XYZ$ (AAS)

d) In the diagram, *PRT* and *SRQ* are straight lines with $PR = RT$ and $SR = RQ$.

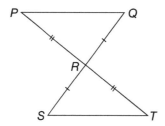

Prove that $\triangle PQR$ is congruent to $\triangle SRT$.

$PR = RT$ (given)
$SR = RQ$ (given)
$\angle PQR = \angle TRS$ (Vertically opposite angles)
So $\triangle PRT \equiv \triangle SRQ$ (SAS)

As well as proving whether triangles are congruent, we can use the congruent rules to find lengths and angles.

Examples

a) Work out the size of each of the angles marked a, b and c.

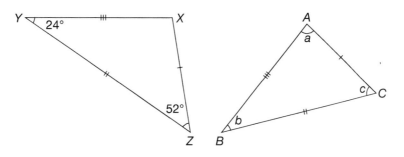

$\angle X = 180° - (52° + 24°)$ (Angles in a triangle add up to 180°.)
    $= 104°$
$\triangle ABC \equiv \triangle XYZ$ (SSS)
So $\angle A = \angle X$, $\angle B = \angle Y$ and $\angle C = \angle Z$
So $a = 104°$, $b = 24°$ and $c = 52°$

b) Work out the size of the angle marked in $\triangle PQR$.

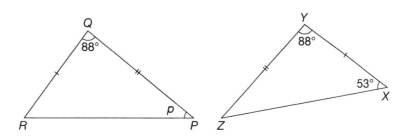

$\triangle PQR \equiv \triangle ZYX$ (SAS)
So $\angle P = \angle Z$
So $p = 180° - (88° + 53°)$ (Angles in a triangle add up to 180°.)
    $= 39°$

c) Work out the size of each of the angles marked a, b, c and d.

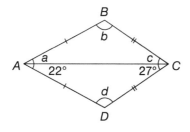

$d = 180° - (22° + 27°)$ (Angles in a triangle add up to 180°.)
    $= 131°$
$\triangle ABC \equiv \triangle ADC$ (SSS)
So $a = 22°$, $b = 131°$, $c = 27°$

d) The line *DE* is parallel to the line *AB* and *AB* = *DE*.

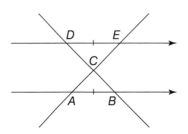

Prove that triangle *ABC* is congruent to triangle *EDC*.

Solution 1

∠*EDC* = ∠*ABC* (alt. ∠s, *AB*||*DE*)
∠*DEC* = ∠*CAB* (alt. ∠s, *AB*||*DE*)
DE = *AB* (given)
So △*ABC* ≡ △*EDC* (ASA)

Solution 2

∠*EDC* = ∠*ABC* (alt. ∠s, *AB*||*DE*)
∠*ACB* = ∠*DCE* (vert. opp. ∠s)
DE = *AB* (given)
So △*ABC* ≡ △*EDC* (AAS)

Can you think of any other ways to prove that these two triangles are congruent?

Exercise 3

1 In each group of triangles, find the pair that are congruent and write down which congruent rule you have used to decide.

HINT: Be careful – the triangles in each group are drawn to look the same even if they are not!

a)

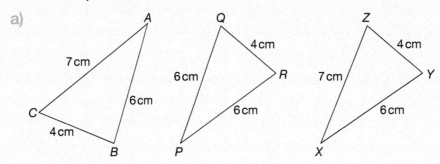

b)

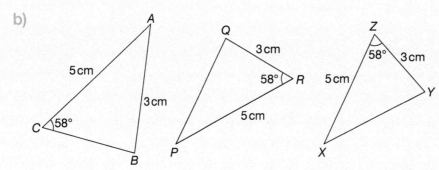

c)

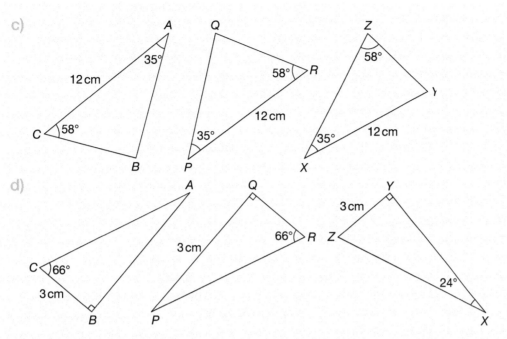

d)

2 Work out the size of each of the angles or sides marked with a lower-case letter. Remember to write down the name of any congruent rule you use.

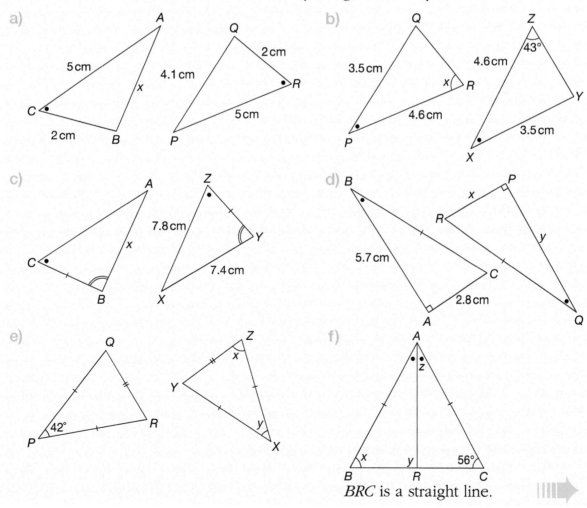

a)

b)

c)

d) B

BRC is a straight line.

e)

f)

g)

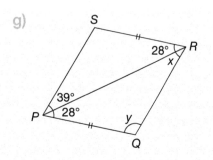

h)

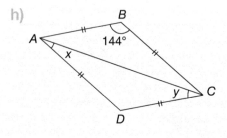

i)

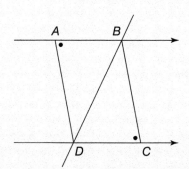

j)

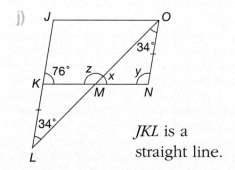

JKL is a
straight line.

3 The line *AB* is parallel to the line *CD*. ∠*BAD* = ∠*DCB*.

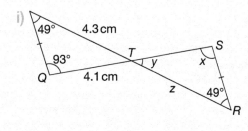

Prove that △*ABD* is congruent to △*CBD*.

4 The line *GC* is parallel to the line *DH*. The line *AE* intersects the line *GC* at
point *B*, and it intersects the line *DH* at point *F*. *AB* = *EF* and ∠*ACB* = ∠*EDF*.

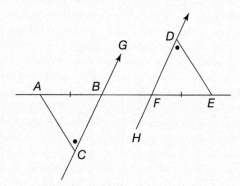

Prove that △*ABC* is congruent to △*EFD*.

D Polygons

We already know that any closed plane figure with three straight sides is called a **triangle** and that any closed plan figure with four straight sides (square, rectangle, rhombus, parallelogram, trapezium and so on) is called a **quadrilateral**.

Closed plane figures with five straight sides or more also have special names. We make these names using the Greek name for the number of sides + 'gon' (which means 'corners' in Greek).

Here are the Greek names for some numbers:

5 = penta   7 = hepta   9 = nona
6 = hexa   8 = octa   10 = deca

So   a closed plane figure with 5 straight sides is called a pentagon
     a closed plane figure with 6 straight sides is called a hexagon
     a closed plane figure with 7 straight sides is called a heptagon
     a closed plane figure with 8 straight sides is called a octagon
     a closed plane figure with 9 straight sides is called a nonagon
and a closed plane figure with 10 straight sides is called a decagon.

We use the word polygons (poly = many) to refer to any many-sided figure – but this doesn't tell us exactly how many sides the figure has.

If all the **sides** in a polygon are the **same length** and all the **angles** are the **same size**, it is a regular polygon.
If the **sides** in a polygon are **different lengths** and/or the **angles** are **different sizes**, it is an irregular polygon.

The sum of the angles inside a polygon

ABCD is any quadrilateral.

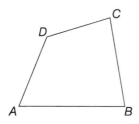

If we join two of the opposite points (either *AC* or *BD*), we make two triangles.

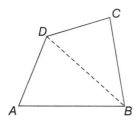

We know that the sum of the angles inside any triangle is 180°.
So the sum of the angles in $\triangle ABD = 180°$
and the sum of the angles in $\triangle BCD = 180°$.
So the sum of the angles in $\triangle ABD + \triangle BCD = 180° + 180° = 360°$.
This tells us that the sum of all the angles inside a quadrilateral is 360°.

Now let's look at a pentagon:

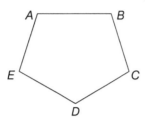

Again it is possible to divide this pentagon into triangles.

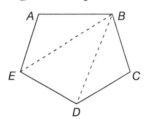

This time we get three triangles.
So the sum of all the angles inside a pentagon is
$180° + 180° + 180° = 540°$.

If we repeat this process for all the other polygons we find these results.

Number of sides	Polygon	Number of triangles	Sum of interior angles
3		1	180°
4		2	$2 \times 180° = 360°$
5		3	$3 \times 180° = 540°$
6		4	$4 \times 180° = 720°$
7		5	$5 \times 180° = 900°$
and so on …	and so on …	and so on …	and so on …
n		$n - 2$	$(n - 2) \times 180°$

For any polygon with n sides, the sum of all interior angles $= (n - 2) \times 180°$.

Examples

a) Calculate the sum of the interior angles of a nonagon.

A nonagon has nine sides, so the sum of the interior
angles $= (9 - 2) \times 180°$
$= 7 \times 180°$
$= 1260°$

b) Calculate the size of each interior angle in a regular octagon.

An octagon has eight sides, so the sum of the interior
angles $= (8 - 2) \times 180°$
$= 6 \times 180°$
$= 1080°$
In a regular octagon, each of the eight interior angles is equal, so
each angle $= 1080° \div 8$
$= 135°$

c) The sum of the interior angles in a polygon is 1980°.
How many sides does this polygon have?

Sum of interior angles of polygon $= (n - 2) \times 180°$
So $(n - 2) \times 180° = 1980°$
$(n - 2) = 1980° \div 180°$
$(n - 2) = 11$
$n = 11 + 2$
$= 13$
So this polygon has 13 sides.

d) In an irregular decagon, nine of the interior angles are equal, and
the last one is a different size.
The nine equal angles are each 138°.
Calculate the size of the other interior angle.

The sum of all the interior angles in a decagon $= (10 - 2) \times 180°$
$= 8 \times 180°$
$= 1440°$
The sum of the nine equal angles $= 9 \times 138°$
$= 1242°$
So the size of the last angle is $1440° - 1242° = 198°$.

Exercise 4

1 Calculate the sum of the interior angles of a polygon with

 a) 11 sides
 b) 15 sides
 c) 20 sides.

2 How many sides does a polygon have if the sum of its interior angles is

 a) 1800°
 b) 2160°
 c) 5040°?

3 How many sides does a regular polygon have if each interior angle is

 a) 140°
 b) 160°
 c) 171°?

4 a) Four of the interior angles of a pentagon are each 110°.
    Calculate the fifth angle.
  b) Four angles in a nonagon are each 140°.
    Calculate the other five angles if they are equal.
  c) Five of the interior angles of a hexagon are 100°, 110°, 125°, 134° and 140°.
    Calculate the size of the sixth interior angle.

5 The sum of the interior angles of a polygon is equal to 12 right-angles.
  Calculate the number of sides in this polygon.

Unit 12 Patterns and sequences

Key vocabulary

connection	order	square numbers
Fibonacci numbers	pattern	term
Fibonacci sequence	related	triangular numbers
linear sequence	sequence	

A What is a sequence?

A pattern can be a special kind of drawing that is made from **repeated** lines, shapes or colours on a surface. We see the same lines, shapes or colours again and again **in an ordered way**, making up the pattern.

The **simplest part** of a pattern that is repeated over and over is just one set of lines, or one shape, or one group of colours. This is called a **unit** (meaning **one**) of the pattern.

Unit	Pattern

We also find patterns in some groups of **numbers**.

For example:

Even numbers	2, 4, 6, 8, 10, …
Odd numbers	1, 3, 5, 7, 9, …
Multiples of 3	3, 6, 9, 12, 15, …
Multiples of 5	5, 10, 15, 20, 25, …
Multiples of 10	10, 20, 30, 40, 50, …

The patterns in these groups of numbers are **not** made in the same way as the other patterns above. We make patterns with lines, shapes and colours by repeating them, but we do not make patterns with numbers by repeating the numbers.

Instead, there is a connection between the numbers, in the order they are written.

For example:
To find the next even number in the group, we add 2 to the last one.
To find the next odd number in the group, we add 2 to the last one.
To find the next multiple of 3 in the group, we add 3 to the last one.
To find the next multiple of 5 in the group, we add 5 to the last one.
To find the next multiple of 10 in the group, we add 10 to the last one.

A group of numbers that are related to each other in the **order** they are written is called a sequence.

> A **sequence** is a list of numbers that is made to follow some **rule**.

Each number in a sequence is called a term (this is the same word as we use for the parts of an expression in algebra).

Each **term** in a sequence has a **special position**. The **order** of the terms is very important – we cannot write them in any order we want. So for each sequence, there will be a special number that is the first term, another special number that is the second term, and so on …

B Using a rule to write the next term in a sequence

Finding the next term given the rule and the first term

If we are given the rule and the first term of a sequence, we can simply apply the rule to the first term to find the second term, and so on until we have as many terms as we need.

Examples

a) The first term of a sequence is 1.
The rule to make the next term is 'multiply the last term by 3, and then add 1'.
Write down the first 5 terms of this sequence.

Apply the rule to the first term:
(first term $\times$ 3) + 1 = (1 $\times$ 3) + 1 = 3 + 1 = 4
Apply the rule to the second term:
(second term $\times$ 3) + 1 = (4 $\times$ 3) + 1 = 12 + 1 = 13
Apply the rule to the third term:
(third term $\times$ 3) + 1 = (13 $\times$ 3) + 1 = 39 + 1 = 40
Apply the rule to the fourth term:
(fourth term $\times$ 3) + 1 = (40 $\times$ 3) + 1 = 120 + 1 = 121
So the first five terms of this sequence are 1, 4, 13, 40 and 121.

b) The first term of a sequence is 5.
The rule to make the next term is 'subtract 3 from the last term, and then multiply by 4'.

i) Write down the first five terms of this sequence.

Apply the rule to the first term:
(first term $-$ 3) $\times$ 4 = (5 $-$ 3) $\times$ 4 = 2 $\times$ 4 = 8
Apply the rule to the second term:
(second term $-$ 3) $\times$ 4 = (8 $-$ 3) $\times$ 4 = 5 $\times$ 4 = 20
Apply the rule to the third term:
(third term $-$ 3) $\times$ 4 = (20 $-$ 3) $\times$ 4 = 17 $\times$ 4 = 68
Apply the rule to the fourth term:
(fourth term $-$ 3) $\times$ 4 = (68 $-$ 3) $\times$ 4 = 65 $\times$ 4 = 260
So the first five terms of this sequence are 5, 8, 20, 68 and 260.

ii) Another sequence is made with the same rule.
This sequence has a different first term.
If the second term is 16, find the first term.

The rule is the same, so (first term $-$ 3) $\times$ 4 = second term
We know the second term is 16, so
(first term $-$ 3) $\times$ 4 = second term
(first term $-$ 3) $\times$ 4 = 16
(first term $-$ 3) = 16 $\div$ 4
first term $-$ 3 = 4
first term = 4 + 3
first term = 7
So the first term of this sequence is 7.

c) The first term of a sequence is 1.
The rule to make the next term is 'add the last two terms'.
Write down the first five terms of this sequence.

Apply the rule to the first term: (first term + 0) = 1 + 0 = 1

NOTE: We add 0 because we have only one term.

Apply the rule to the second term:
(second term + first term) = 1 + 1 = 2
Apply the rule to the third term:
(third term + second term) = 2 + 1 = 3
Apply the rule to the fourth term:
(fourth term + third term) = 3 + 2 = 5
So the first five terms of this sequence are 1, 1, 2, 3 and 5.
This is a very special sequence called the Fibonacci sequence.
We will learn more about this later.
Can you write down the next four terms in this sequence?

Finding the rule

If we are given at least three terms in a sequence, it is usually
possible to work out the rule to find the next term in the sequence.

Examples

For each of these sequences:
 i) write down the rule to make the next term in words
ii) write down the next four terms.

a) 1, 3, 9, 27 …
   i) Each term is three times bigger than the previous term.
      So the rule is 'multiply the last term by 3'.
   ii) Applying this rule, the next four terms will be:
        $27 \times 3 = 81$
        $81 \times 3 = 243$
       $243 \times 3 = 729$
       $729 \times 3 = 2187$

b) 256, 64, 16, 4 …
   i) Each term is four times smaller than the previous term.
      So the rule is 'divide the last term by 4'.
   ii) Applying this rule, the next four terms will be:
        $4 \div 4 = 1$
        $1 \div 4 = \frac{1}{4}$ or 0.25
        $\frac{1}{4} \div 4 = \frac{1}{16}$ or 0.0625
        $\frac{1}{16} \div 4 = \frac{1}{64}$ or 0.015625

c) 19, 14, 9, 4 …

    i) Each term is five smaller than the previous term.
So the rule is 'subtract 5 from the last term'.

    ii) Applying this rule, the next four terms will be:
$$4 - 5 = -1$$
$$-1 - 5 = -6$$
$$-6 - 5 = -11$$
$$-11 - 5 = -16$$

Look back at the examples above. They show that:
    sequences can get bigger or smaller
    sequences can include negative numbers
    sequences can include fractions (or decimal fractions).

Exercise 1

1 Write down the next three terms in each of these sequences.

a) 9, 13, 17, 21, …
b) 10, 12, 14, 16, …
c) 28, 25, 22, 19, …

d) 13, 18, 23, 28, 33, …
e) 3, 6, 12, 24, …
f) $\frac{1}{4}, \frac{1}{2}, \frac{3}{4}, 1, 1\frac{1}{4}$, …

g) 32, 16, 8, 4, …
h) 0.7, 0.8, 0.9, 1.0, …
i) 10, 8, 6, 4, …

j) 120, 60, 30, …
k) 1, 3, 6, 10, 15, …
l) 1, 3, 4, 7, 11, 18, …

2 Write down the missing terms for each of these sequences.

a) 2, 4, 6, ?, 10, 12, ?, 16, …
b) 2, 6, ?, 14, 18, ?, 26, …

c) 1, 2, 4, ?, 16, ?, 64, …
d) 28, 22, ?, 10, 4, ?, …

e) 1, 4, 9, ?, 25, ?, 49, …
f) 1, 2, 3, 5, ?, 13, ?, 34, …

g) ?, 8, 14, ?, ?, 32, 38, …
h) ?, 1.8, ?, 1.4, ?, 1, 0.8, …

3 For each of these sequences:
  i) write down the next two terms
  ii) write down the rule to make the next term in words.
For example:

Sequence   **Rule in words**
5, 4, 3, 2, 1, 0, −1   *Subtract 1 from the last term.*

a) 2, 9, 16, 23, 30, …
b) 3, 5, 7, 9, 11, …

c) 1, 5, 9, 13, 17, …
d) 31, 26, 21, 16, …

e) 64, 32, 16, 8, 4, …
f) 1, 3, 9, 27, …

g) 0.2, 0.4, 0.6, 0.8, …
h) $\frac{3}{4}, 1\frac{1}{2}, 2\frac{1}{4}, 3$, …

i) 4, 6, 10, 16, 24, …
j) 10, 7, 4, 1, −2, …

4 A sequence is made using the rule 'subtract 3 from the last term, and then divide by 2'.

　a) If the first term is 45, write down the second, third and fourth terms.
　b) A different sequence is written using the same rule.
　　The second term of this sequence is 17.
　　What is the first term of this sequence?

5 A sequence is made using the rule 'add 5 to the last term, and then multiply by 3'.

　a) If the first term is 7, write down the second, third and fourth terms.
　b) A different sequence is written using the same rule.
　　The second term of this sequence is 45.
　　What is the first term of this sequence?

6 A sequence is made using the rule 'add the last two terms, and then multiply by 3'.

　If the first two terms are 1, −3, write down the third, fourth and fifth terms.

7 A sequence is made using the rule 'multiply the last term by 2, and then subtract 1'.

　a) If the first term is 4, write down the second and third terms.
　b) The 11th term in this sequence is 3073. Use this information to work out the 10th term.

 # Using a rule to write any term in a linear sequence

So far, we have learned about rules that help us to write the **next term** in a sequence. This is fine for a few terms, but gets more difficult later on – if we have the first five terms of a sequence and the rule to write the next term, it will be a lot of work to find the 100th term.

It is much quicker and easier to use a rule that tells us how to write **any** term in a sequence. This kind of rule will be in the form of an algebraic equation.

For example, any term in one sequence can be written with the rule $T = 4n - 1$.
The variable T stands for the term.
The variable n stands for the number of the position this term has in the sequence (so for the first term $n = 1$, for the second term $n = 2$, for the fifteenth term $n = 15$, and so on).

Now it is easy to find any term in the sequence!

Example

Find the 100th term of the sequence defined by the rule $T = 4n - 1$.

For the 100th term, $n = 100$. Substitute this in the rule.

$$T = 4n - 1$$
$$\text{the 100th term} = (4 \times 100) - 1$$
$$= 400 - 1$$
$$= 399$$

D Finding a rule to write any term in a linear sequence

If we are given some terms of a sequence, we can find the rule to write any term in a linear sequence by looking at the difference between the terms.

Example

Look at this sequence: 3, 6, 9, 12, 15, …
Find the rule to write any term in this sequence.

First we need to work out the differences between the terms.

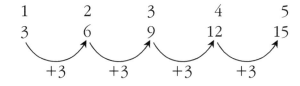

number of the term, n	1	2	3	4	5
term, T	3	6	9	12	15
difference		+3	+3	+3	+3

The differences are all $+3$. This means that every time the term number, n, goes up by 1, the value of the term, T, goes up by 3. So, the value of T is always $(+3)$ times the value of n. In other words, $T = (+3)n$.

It is important to check that this is correct for the terms that we know:

First term: $T = (+3)n = (+3) \times 1 = 3$ ✓
Second term: $T = (+3)n = (+3) \times 2 = 6$ ✓
Third term: $T = (+3)n = (+3) \times 3 = 9$ ✓
Fourth term: $T = (+3)n = (+3) \times 4 = 12$ ✓
Fifth term: $T = (+3)n = (+3) \times 5 = 15$ ✓

Using the rule to write the **next term**, the 6th term is
last term $+ 3 = 15 + 3 = 18$.
Using our new rule to write **any term**, the 6th term is
$T = (+3)n = (+3) \times 6 = +18$.
Both methods give the same answer, so our rule is correct.
The rule to write any term in this sequence is $T = 3n$.

Sometimes we have to do more than just multiply or divide the number of the term – we may need to add or subtract a constant as well.

Examples

a) Look at this sequence: 5, 8, 11, 14, 17, …
   Find the rule to write any term in this sequence.

Work out the differences between the terms.

number of the term, n	1	2	3	4	5
term, T	5	8	11	14	17
difference		+3	+3	+3	+3

Once again the differences are all +3. This means that every time the term number, n, goes up by 1, the value of the term, T, goes up by 3.
But this time the value of T is not exactly (+3) times the value of n.
So now we must compare the values of $3n$ with the value of T to see what we must add or subtract.

first term	$n = 1$	$T = 5$	$3n = (3 \times 1) = 3$	So we must add 2, giving $T = 3n + 2$
second term	$n = 2$	$T = 8$	$3n = (3 \times 2) = 6$	So we must add 2, giving $T = 3n + 2$
third term	$n = 3$	$T = 11$	$3n = (3 \times 3) = 9$	So we must add 2, giving $T = 3n + 2$
fourth term	$n = 4$	$T = 14$	$3n = (3 \times 4) = 12$	So we must add 2, giving $T = 3n + 2$
fifth term	$n = 5$	$T = 17$	$3n = (3 \times 5) = 15$	So we must add 2, giving $T = 3n + 2$

This gives us the same rule for every term.
So for this sequence, the rule to write any term is. $T = 3n + 2$.

b) Look at this sequence: −1, 3, 7, 11, 15, …
   Find the rule to write any term in this sequence.

Work out the differences between the terms.

number of the term, n	1	2	3	4	5
term, T	−1	3	7	11	15
difference		+4	+4	+4	+4

The differences are all +4. This means that every time the term number, n, goes up by 1, the value of the term, T, goes up by 4.
But the value of T is not exactly (+4) times the value of n.
We must compare the values of $4n$ with the value of T to see what we must add or subtract.

first term $n = 1$ $T = -1$ $4n = (4 \times 1) = 4$ So we must subtract 5, giving $T = 4n - 5$
second term $n = 2$ $T = 3$ $4n = (4 \times 2) = 8$ So we must subtract 5, giving $T = 4n - 5$
third term $n = 3$ $T = 7$ $4n = (4 \times 3) = 12$ So we must subtract 5, giving $T = 4n - 5$
fourth term $n = 4$ $T = 11$ $4n = (4 \times 4) = 16$ So we must subtract 5, giving $T = 4n - 5$
fifth term $n = 5$ $T = 15$ $4n = (4 \times 5) = 20$ So we must subtract 5, giving $T = 4n - 5$

This gives us the same rule for every term.
So for this sequence, the rule to write any term is $T = 4n - 5$.

c) Look at this sequence: 16, 13, 10, 7, 4, …
Find the rule to write any term in this sequence.

Work out the differences between the terms.

number of the term, n	1	2	3	4	5
term, T	16	13	10	7	4
difference		-3	-3	-3	-3

The differences are all -3. This means that every time the term number, n, goes up by 1, the value of the term, T, goes down by 3. But the value of T is not exactly (-3) times the value of n. We must compare the values of $-3n$ with the value of T to see what we must add or subtract.

first term $n = 1$ $T = 16$ $-3n = (-3 \times 1) = -3$ So we must add 19, giving $T = -3n + 19$
second term $n = 2$ $T = 13$ $-3n = (-3 \times 2) = -6$ So we must add 19, giving $T = -3n + 19$
third term $n = 3$ $T = 10$ $-3n = (-3 \times 3) = -9$ So we must add 19, giving $T = -3n + 19$
fourth term $n = 4$ $T = 7$ $-3n = (-3 \times 4) = -12$ So we must add 19, giving $T = -3n + 19$
fifth term $n = 5$ $T = 4$ $-3n = (-3 \times 5) = -15$ So we must add 19, giving $T = -3n + 19$

This gives us the same rule for every term.
So for this sequence, the rule to write any term is $T = -3n + 19$.

A number sequence that increases (or decreases) by the same amount from one term to the next term is called a linear sequence.

To find the rule for **any term** in a linear sequence we use these simple steps:

1 Calculate the **difference** between each term and the next one in the sequence.
2 If this difference is the same for all terms, the sequence is **linear**.
3 Calculate (difference $\times n$) for each of the terms.
4 Calculate the number you must add to or subtract from (difference $\times n$) to make T for each term.
   Call this number $+c$ if we must add it and $-c$ if we must subtract it.
5 The rule will be $T = $ (difference $\times n$) $\pm c$.

Exercise 2

1 The rule to write any term in a sequence is $T = 2n + 1$.
 a) Work out the first five terms in this sequence.
 b) Work out the 25th term.

2 The rule to write any term in a sequence is $T = 5n - 2$.
 a) Work out the first five terms in this sequence.
 b) Work out the 30th term.

3 The rule to write any term in a sequence is $T = n - 10$.
 a) Work out the first five terms in this sequence.
 b) Work out the 322nd term.

4 The rule to write any term in a sequence is $T = -n + 5$.
 a) Work out the first five terms in this sequence.
 b) Work out the 97th term.

5 The rule to write any term in a sequence is $T = 8 - 2n$.
 a) Work out the first five terms in this sequence.
 b) Work out the 150th term.

6 The rule to write any term in a sequence is $T = -4n - 7$.
 a) Work out the first five terms in this sequence.
 b) Work out the 250th term.

7 A sequence of numbers starts: 4, 7, 10, 13, …
 a) What is the difference between terms?
 b) Is this sequence linear?
 c) Work out the rule to find any term in this sequence.
 d) Work out the 25th term.
 e) Work out the 110th term.

8 A sequence of numbers starts: 9, 11, 13, 15, …
 a) What is the difference between terms?
 b) Is this sequence linear?
 c) Work out the rule to find any term in this sequence.
 d) Work out the 25th term.
 e) Work out the 110th term.

9 A sequence of numbers starts: 20, 16, 12, 8, …
 a) What is the difference between terms?
 b) Is this sequence linear?
 c) Work out the rule to find any term in this sequence.
 d) Work out the 25th term.
 e) Work out the 110th term.

10 A sequence of numbers starts: 5, 9, 13, 17, …

   a) What is the difference between terms?
   b) Is this sequence linear?
   c) Work out the rule to find any term in this sequence.
   d) Work out the 25th term.
   e) Work out the 110th term.

11 A sequence of numbers starts: 19, 16, 13, 10, …

   a) What is the difference between terms?
   b) Is this sequence linear?
   c) Work out the rule to find any term in this sequence.
   d) Work out the 25th term.
   e) Work out the 110th term.

12 A sequence of numbers starts: 4, 8, 12, 16, …

   a) What is the difference between terms?
   b) Is this sequence linear?
   c) Work out the rule to find any term in this sequence.
   d) Work out the 25th term.
   e) Work out the 110th term.

13 A sequence of numbers starts: 1, 3, 5, 7, …

   a) What is the difference between terms?
   b) Is this sequence linear?
   c) Work out the rule to find any term in this sequence.
   d) Work out the 25th term.
   e) Work out the 110th term.

14 A sequence of numbers starts: 7, 11, 15, 19, …

   a) What is the difference between terms?
   b) Is this sequence linear?
   c) Work out the rule to find any term in this sequence.
   d) Work out the 25th term.
   e) Work out the 110th term.

15 A sequence of numbers starts: 6, 4, 2, 0, …

   a) What is the difference between terms?
   b) Is this sequence linear?
   c) Work out the rule to find any term in this sequence.
   d) Work out the 25th term.
   e) Work out the 110th term.

16 A sequence of numbers starts: 3, 8, 13, 18, …

　a) What is the difference between terms?

　b) Is this sequence linear?

　c) Work out the rule to find any term in this sequence.

　d) Work out the 25th term.

　e) Work out the 110th term.

17 A sequence of numbers starts: 40, 35, 30, 25, …

　a) What is the difference between terms?

　b) Is this sequence linear?

　c) Work out the rule to find any term in this sequence.

　d) Work out the 25th term.

　e) Work out the 110th term.

18 A sequence of numbers starts: 0, 1, 2, 3, …

　a) What is the difference between terms?

　b) Is this sequence linear?

　c) Work out the rule to find any term in this sequence.

　d) Work out the 25th term.

　e) Work out the 110th term.

19 A sequence of numbers starts: −1, 1, 3, 5, …

　a) What is the difference between terms?

　b) Is this sequence linear?

　c) Work out the rule to find any term in this sequence.

　d) Work out the 25th term.

　e) Work out the 110th term.

20 A sequence of numbers starts: 1.5, 2, 2.5, 3, …

　a) What is the difference between terms?

　b) Is this sequence linear?

　c) Work out the rule to find any term in this sequence.

　d) Work out the 25th term.

　e) Work out the 110th term.

21 A sequence of numbers starts: 2.25, 2, 1.75, 1.5, 1.25, …

　a) What is the difference between terms?

　b) Is this sequence linear?

　c) Work out the rule to find any term in this sequence.

　d) Work out the 25th term.

　e) Work out the 110th term.

Number sequences from shape patterns

A group of patterns made from the same unit shape can be changed into a number sequence.

All we need to do is count the number of unit shapes in each pattern, then write the number of units in each pattern in the same order to make a number sequence.

From the number sequence, we can work out the rule to write any term, as we did before.

Examples

Each of these patterns is made from the unit shape of a square.

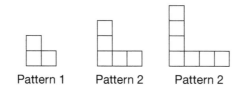

Pattern 1   Pattern 2   Pattern 2

a) How many units will make up pattern 4?

Pattern 1 is made from 3 units.
Pattern 2 is made form 5 units.
Pattern 3 is made from 7 units.
We can write this as a number sequence: 3, 5, 7, …
By looking at either the patterns or the number sequence, we can tell that pattern 4 will have 9 units.

b) Find the rule to write any term in the sequence represented by these patterns.

Calculate the difference between terms: it is $+2$.
The differences are all the same, so this is a linear sequence.
Calculate (difference $\times n$) for each of the terms and find c.

pattern 1 $n = 1$ $T = 3$ $2n = 2 \times 1 = 2$ So we must add 1, giving $T = 2n + 1$
pattern 2 $n = 2$ $T = 5$ $2n = 2 \times 2 = 4$ So we must add 1, giving $T = 2n + 1$
pattern 3 $n = 3$ $T = 7$ $2n = 2 \times 3 = 6$ So we must add 1, giving $T = 2n + 1$
pattern 4 $n = 4$ $T = 9$ $2n = 2 \times 4 = 8$ So we must add 1, giving $T = 2n + 1$

The rule for this sequence is $T = 2n + 1$.

c) Draw the 15th pattern.

The 15th term is $T = 2n + 1 = (2 \times 15) + 1 = 31$
The pattern would look like this.

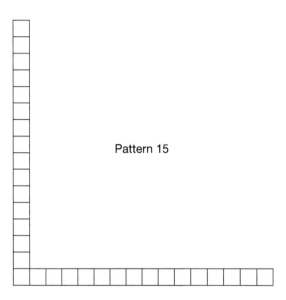

Pattern 15

Can you explain why none of these patterns will have an even number of unit shapes?

 # Some special number sequences

Fibonacci numbers

This famous set of numbers forms a sequence that is used to help work out many different things in mathematics.

It is named after Leonardo of Pisa, who was also called Fibonacci (don't confuse him with the more famous Leonardo da Vinci, who lived more than 200 years later).

This special sequence of numbers can be found in nature – in the branching of trees, or the arrangement of a pine cone. Fibonacci numbers are also sometimes used in music to work out tunings and sometimes in art as well.

Each term in this sequence is made by adding together the **last two terms**.

Sometimes the sequence is written with 0 as the first term, but often this is assumed and the sequence is written as:
1, 1, 2, 3, 5, 8, 13, 21, 34, 55, …

A French mathematician, Blaise Pascal, developed a triangle that is based on the Fibonacci numbers – called Pascal's Triangle. Write in some more of the numbers yourself.

							1							
						1		1						
					1		2		1					
				1		3		3		1				
			1		4		6		4		1			
		1		5		10		10		5		1		
	1		6		15		20		15		6		1	
1		7		21		35		35		21		7		1
1	8	28	56	70	56	28	8	1						

Each number in this triangle is the sum of the two numbers diagonally above it on the right and the left.

This triangle has many very interesting and unusual properties. It can be used for many things:

- Horizontal rows add to powers of 2 (1, 2, 4, 8, 16, …).
- If we read the horizontal rows as single numbers, they are the powers of 11 (1, 11, 121, 1331, 14 641, …).
- Adding any two successive numbers in the diagonal 1-3-6-10-15-21-28... results in a perfect square (1, 4, 9, 16, …).
- It can be used to find combinations in probability problems. For example, if you pick any **two** of **five** items, the number of possible combinations is 10 – this can be found in the triangle by looking in the **second** place of the **fifth** row (do not count the 1s).
- When the first number to the right of the 1 in any row is a prime number, all numbers in that row are divisible by that prime number.

Square numbers

This is the sequence of numbers that are the **exact squares** of the counting numbers. It is made with the numbers 1^2, 2^2, 3^2, 4^2, … and is written as 1, 4, 9, 16, 25, 36, 49, 64, …

These numbers are called the square numbers because they make exactly square patterns.

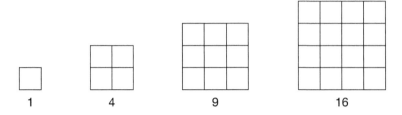

1   4   9   16

Triangular numbers

This sequence of numbers starts like this: 1, 3, 6, 10, 15, 21, 28, 36, …

These numbers are called the triangular numbers because they make triangular patterns.

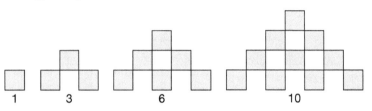

1   3   6   10

Exercise 3

1

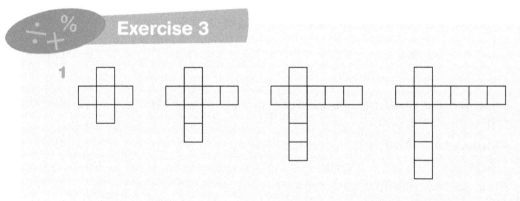

a) Write down the sequence of numbers from these patterns.
b) Work out the rule to write any term in this sequence.
c) How many squares will make the 50th pattern?

2

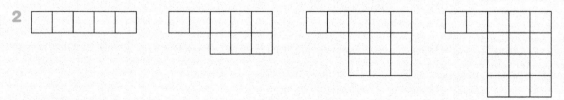

a) Write down the sequence of numbers from these patterns.
b) Work out the rule to write any term in this sequence.
c) Which pattern will use 173 squares?

3 These patterns are made using equilateral triangles.

  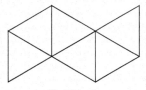 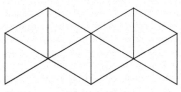

Pattern 1   Pattern 2   Pattern 3   Pattern 4

a) Write down the sequence of numbers from these patterns.
b) Work out the rule to write any term in this sequence.
c) Why will a pattern in this sequence never have 27 triangles?

4 These patterns are made using matchsticks.

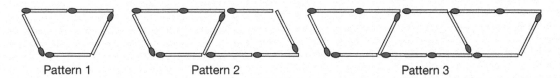

Pattern 1   Pattern 2   Pattern 3

a) Write down the sequence of numbers from these patterns.
b) Work out the rule to write any term in this sequence.
c) Which pattern will use 93 matchsticks?

5 Ranch fences are made by placing fence poles 1 m apart.
Two horizontal bars are placed between each two fence poles.
The fence in the picture below is 4 m long. It has five fence poles and eight horizontal bars.

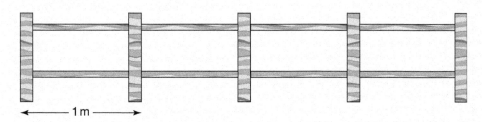

← 1 m →

a) Copy and complete this table about the fence.

Metres	1	2	3	4	5
Number of bars					
Number of poles					

b) The number of bars forms a number sequence.
Write a rule for the number of bars for n metres.
c) The number of poles forms a different number sequence.
Write a rule for the number of poles for n metres.

d) If the fence is 50 m long, calculate:
   i) the number of bars
   ii) the number of poles.

6 Cubes of side 1 cm are joined together to make square poles.
This pole is made with four cubes linked together.

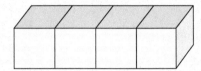

The surface area of the pole is 18 cm^2 (four squares on each of the long sides plus one square at each end).

a) What is the surface area of a pole made using five cubes?
b) What is the surface area of a pole made using ten cubes?
c) What is the surface area of a pole made using n cubes?

7 These are the first four triangular numbers: 1, 3, 6, 10, …

a) What is the next triangular number?
b) Copy and complete this table for the first five triangular numbers.

$\dfrac{1 \times 2}{2} = 1$	$\dfrac{2 \times 3}{2} = 3$	$\dfrac{3 \times 4}{2} = 6$	$\dfrac{}{2} = 10$	$\dfrac{}{} =$

c) Write a rule for the nth triangular number.
d) Calculate the 10th triangular number.

Presenting statistical data

Key vocabulary

bar chart	histogram	sector
column graph	line graph	statistics
data	pictogram	stem-and-leaf diagram
frequency table	pie chart	symbol

A Data and statistics

Information (facts and figures) from real life is called data.

Every day, we see statements that give us facts and figures about many different things. These are all examples of data:

- 345 pupils applied to study on the English Programme at a school in 2008. 69 pupils passed the application test and interview. This is a 20% pass rate.
- There are 127 secondary schools in Chiang Rai.
- 9 out of 10 pop stars use this shampoo.
- There were 153 deaths resulting from drowning in the floods of 2006.

Statistics is the science of collecting, organising, interpreting and analysing data.

In Coursebook 3 we will learn more about statistics, but for now we are going to look at the different ways we can use to **present** data (show it to people).

If there is a lot of data, it is often difficult to understand when it is just a list of numbers. It is usually easier to understand things when we see some kind of **picture**.

A **graph** is a 'picture' that shows us the information from a set of data. It is easier to **compare** different parts of the data and see how they are **connected** when we can look at a graph rather than a long list of numbers.

There are many different types of graph we can use to present statistical data:

- Line graphs
- Pictograms
- Bar charts
- Frequency tables
- Histograms
- Stem-and-leaf diagrams
- Pie charts

B Line graphs

We have already learned about drawing line graphs. The data is plotted on a Cartesian plane using x- and y-coordinates.

Example

The temperature is recorded every two hours for one day.
Here is the data.

Time	02.00	04.00	06.00	08.00	10.00	12.00	14.00	16.00	18.00	20.00	22.00
Temperature in °C	10	8	6	9	15	17	20	22	16	13	11

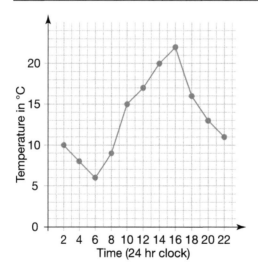

C Pictograms

In a pictogram (sometimes called a **pictograph**), some sort of picture symbol is used to stand for a chosen number of items. This is called the **scale**.

Example   If we use the symbol ♪ to stand for four pupils, a pictogram of the numbers of pupils studying a musical instrument in each year at a school would look like this.

Number of pupils studying a musical instrument

Year 7	♪ ♪ ♪ ♪ ♪ ♪ ♪ ♪ ♪
Year 8	♪ ♪ ♪ ♪ ♪
Year 9	♪ ♪ ♪ ♪ ♪ ♪ ♪
Year 10	♪ ♪ ♪ ♪
Year 11	♪ ♪
Year 12	♪ ♪ ♪

Scale: ♪ stands for four pupils

Pictograms can look fun, but they are not often useful in practice as there are some problems.

- If we use 👤 to stand for 10 people, we could draw half a man to show five people – but how will we show eight people, and so on … ?
- Drawing pictograms usually takes quite a long time because the symbols have to be drawn one by one and exactly the same.

D   Bar charts

In a pictogram we use rows of picture symbols to stand for different numbers of each item. In a bar chart (sometimes called a **bar graph**) we use simple plain 'bars' of different lengths to stand for different numbers of each item.

A bar chart can be drawn with **horizontal bars** or with **vertical bars**.

Example   The pictogram above can be changed into a bar chart showing the same data.

Year 7	♪ ♪ ♪ ♪ ♪ ♪ ♪ ♪ ♪
Year 8	♪ ♪ ♪ ♪ ♪
Year 9	♪ ♪ ♪ ♪ ♪ ♪ ♪
Year 10	♪ ♪ ♪ ♪
Year 11	♪ ♪
Year 12	♪ ♪ ♪

➡

Year 7	♪ ♪ ♪ ♪ ♪ ♪ ♪ ♪ ♪
Year 8	♪ ♪ ♪ ♪ ♪
Year 9	♪ ♪ ♪ ♪ ♪ ♪ ♪
Year 10	♪ ♪ ♪ ♪
Year 11	♪ ♪
Year 12	♪ ♪ ♪

We now need a scale along the x-axis so that we know the number of pupils shown by each of the bars.

Number of pupils studying a musical instrument

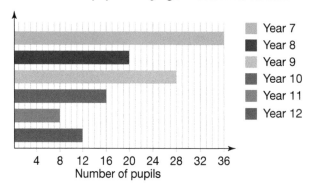

If we use vertical bars, the chart is often called a column graph.

Number of pupils studying a musical instrument

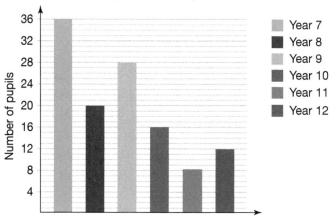

E Frequency tables

'**Frequency**' means 'how often' something happens.

We can use a frequency table to record the results of an experiment **as it is happening**. Each time something happens, we count it by drawing a small stroke in the table. We draw the fifth stroke **across** the other four so that it is easier to count up the total at the end. So a frequency of 7 looks like this: 卌 ||

Example

A pupil counts the colours of the cars at an intersection for half an hour. Here are his results in the form of a frequency table.

Colour	Frequency	Total
White	卌 卌 卌 ||||	19
Black	卌 卌 卌 卌 卌 |||	28
Silver	卌 卌 卌 卌 卌 卌 卌		36
Gold	卌 卌 卌 卌 卌 ||	27

F Histograms

A histogram is really just a **bar chart**, but the data shown in a histogram is **always** from a **frequency table**.

Example   The histogram for the data in the frequency table on page 248 looks like this.

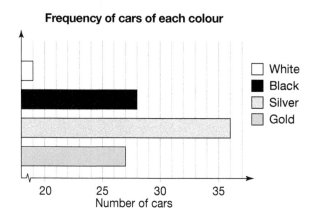

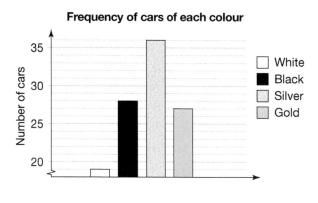

G Stem-and-leaf diagrams

Stem-and-leaf diagrams are useful to summarise a set of data without losing the detail. These diagrams are also helpful in finding **patterns** or **very extreme data scores**.

To make a stem-and-leaf diagram we break the numbers of each score into two parts. The left group of 'leading' numbers is called a **stem**, and the other group of 'trailing' numbers on the right is called a **leaf**.

Example   Here is a list of weights (in kg) of a group of school children.

Girls: 41 37 44 46 29 37 39 42 51 38
Boys: 31 34 36 33 43 38 26 38 41 39

We can see that the weights are in 20s, 30s, 40s and 50s. So we can use the first number in each weight as the 'stem' and the second number as the 'leaf'.
First write out the stem:

```
2 |
3 |
4 |
5 |
```

The first number is 41:

```
2 |
3 |
4 | 1
5 |
```

The second number is 37:

```
2 |
3 | 7
4 | 1
5 |
```

Continue adding the numbers until they have all been recorded:

Weights (in kg) of some school children

```
2 | 9 6
3 | 7 7 9 8 1 4 6 3 8 8 9
4 | 1 4 6 2 3 1
5 | 1
```
Key: 2 | 9 = 29 kg

If we don't have a large amount of data, we can leave the stem-and-leaf diagram like this, **unordered**. If we do have a large amount of data, an **ordered** diagram might be clearer. This simply means that we rewrite the diagram with the numbers in each leaf in order.

Weights (in kg) of some school children

```
2 | 6 9
3 | 1 3 4 6 7 7 8 8 8 9 9
4 | 1 1 2 3 4 6
5 | 1
```
Key: 2 | 9 = 29 kg

We can also separate the 'leafs' for the boys and girls.

Weights (in kg) of some school children

```
Girls              Boys
        9 | 2 | 6
    9 8 7 7 | 3 | 1 3 4 6 8 8 9
    6 4 2 1 | 4 | 1 3
          1 | 5 |
```
Key: 2 | 9 = 29 kg

'Leafs' shown as single digits like this are usually easier to read and interpret than the original data.

 Pie charts

The pie chart is different from other types of graphs. To draw a pie chart we do not need graph paper with a grid of squares on it. It also does not need a scale.

Each category of data is shown by a 'slice' of the pie chart. The pie chart is a 'picture' that allows us easily to compare the different data scores with each other – and also their fractional or percentage part of the whole.

A 'slice' of a circle is called a 'sector'. So each of the data scores is shown by the **area** of one of the sectors in the pie chart. This means that each data score is **proportional to the area of the sector** showing that score.

$$\frac{\text{Area of sector}}{\text{Area of circle}} = \frac{\text{Angle of sector}}{360°}$$

So the **angle** of a sector is also proportional to the area of that sector. This means that we can use the angle of a sector to stand for each of the data scores.

Example

Some school pupils were asked how they come to school each day. Here is the data from their answers.

Type of transport	Number of pupils
Bus	13
Walk	10
Car	6
Cycle	11

The **total** number of pupils in this set of data is
13 + 10 + 6 + 11 = 40
A **full circle** has 360°.
We will use a full circle to stand for the total of 40 pupils.

13 pupils out of 40 is a fraction of $\frac{13}{40}$. This fraction of the full circle is $\frac{13}{40} \times 360° = 117°$.
So the angle of the sector to show the pupils who travel by bus is 117°.

It often helps to set out pie chart calculations in a table.

Type of transport	Number of pupils	Fractional part	Angle of sector
Bus	13	$\dfrac{13}{40}$	$\dfrac{13}{40} \times 360° = 117°$
Walk	10	$\dfrac{10}{40}$	$\dfrac{10}{40} \times 360° = 90°$
Car	6	$\dfrac{6}{40}$	$\dfrac{6}{40} \times 360° = 54°$
Cycle	11	$\dfrac{11}{40}$	$\dfrac{11}{40} \times 360° = 99°$
TOTAL	40	$\dfrac{40}{40} = 1$	$= 360°$

Check that the total of the angles of all the sectors adds up to 360°.

Now we draw a large circle with a pair of **compasses**. We use a **protractor** to measure the angles for each sector.

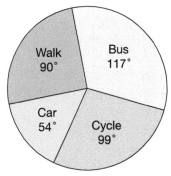

Types of transport used by some pupils

Always **label** the sectors of the pie chart.

Normally, when we draw a pie chart to show data to the public, we do not write in the size of the sector angles. However, for all of the exercises in this section of your maths study, you should write in the angles.

Sometimes we are given a pie chart and need to work out some information from it.

Here is a pie chart. It shows the different things that are in some food.

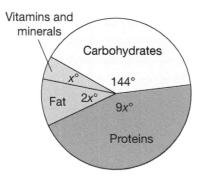

Vitamins and minerals

Carbohydrates

$x°$

$144°$

Fat $2x°$

$9x°$

Proteins

a) Find the value of x.

$$9x + 2x + x + 144° = 360°$$
$$12x = 216°$$
$$x = 18°$$

b) What is the percentage of fat in this food?

The angle in the pie chart that stands for the amount of fat is $2x° = 2 \times 18° = 36°$.
So, the fraction of the whole circle used to show the fat is
$\frac{36°}{360°} = \frac{1}{10}$.
This means that $\frac{1}{10}$ of the food is fat. As a percentage, this is $\frac{1}{10} \times 100\% = 10\%$.

c) If this food contains 120 g of carbohydrates, calculate the total weight of the food.

The angle in the pie chart that stands for the amount of carbohydrate is 144°.
We know that 144° = 120 g and that 360° stands for the whole amount of the food.
So $360° = \left(\frac{120}{144} \times 360\right)$
$\qquad = 300\,g$

1 This pictogram shows the number of trees of different types in Lincoln Park.

Beech	🌳 🌳 🌳
Oak	🌳 🌳 🌳 🌳 🌳
London plane	🌳 🌳 🌳 🌳 🌳 🌳 🌳 🌳 🌳
Chestnut	🌳 🌳
Maple	🌳

Scale: 🌳 stands for four trees

a) What is the most common type of tree in Lincoln Park?

b) What is the least common type of tree in Lincoln Park?

c) Write down the exact number of each type of tree in Lincoln Park.

d) Two more chestnut trees and six more beech trees are planted.
Draw a pictogram to show the new numbers of each type of tree in Lincoln Park.

2 Draw a column graph to show the information given in the pictogram in question 1.

3 Some pupils counted the number of vehicles passing an intersection between 7 a.m. and 5 p.m. on one day. Their data is shown in this column graph.

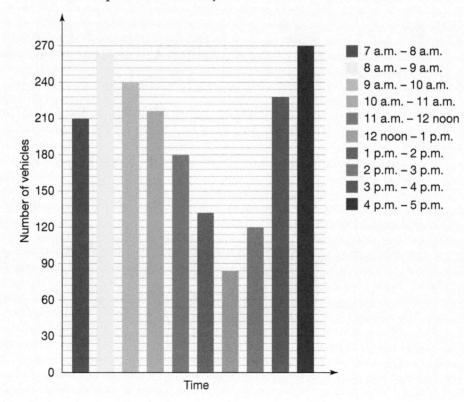

■	7 a.m. – 8 a.m.
▢	8 a.m. – 9 a.m.
▨	9 a.m. – 10 a.m.
▨	10 a.m. – 11 a.m.
▨	11 a.m. – 12 noon
▨	12 noon – 1 p.m.
■	1 p.m. – 2 p.m.
▨	2 p.m. – 3 p.m.
■	3 p.m. – 4 p.m.
■	4 p.m. – 5 p.m.

a) When is this intersection busiest in the morning?

b) When is this intersection busiest in the afternoon?

c) When is this intersection the quietest during the day?

d) How many vehicles does each small division on the vertical axis stand for?

e) How many vehicles passed the intersection between

   i) 9 a.m. and 11 a.m.

   ii) 12 noon and 2 p.m.

   iii) 3 p.m. and 5 p.m.?

f) During which hour did these numbers of vehicles pass the intersection?

   i) 228

   ii) 132

   ii) 264

g) How many vehicles passed the intersection between 10 a.m. and 2 p.m.?

h) How many more vehicles passed the intersection between 7 a.m. and 8 a.m. than between 2 p.m. and 3 p.m.?

4 A person who works in a customer service call centre keeps a record of the number of minutes each telephone call lasts. Here are her results.

11	26	32	40	20
14	27	9	10	11
16	21	27	20	27
10	14	31	29	34

a) Draw a stem-and-leaf diagram to show this set of data.

b) How many minutes does the longest call last?

c) How many minutes does the shortest call last?

5 Here is a list of the systolic blood pressures of some people.

160	131	148	151	154
183	144	161	150	154
129	166	176	206	160
129	175	159	137	151
123	135	158	189	198
185	153	132	180	170

a) Draw a stem-and-leaf diagram with one-digit leaves to show this set of data.

b) How many people's blood pressure was measured?

c) What is the lowest systolic pressure measured?

d) What is the highest systolic pressure measured?

6 The families in a small village have a number of different pets. The data is set out in this table.

Pet	Number	Fractional part	Angle of sector
Cats	36		
Dogs	20		
Birds	6		
Goats	4		
Rabbits	14		
TOTAL			= 360°

Copy the table and complete the calculations.
Then draw a pie chart to show this information.

7 At 6 p.m. last Thursday, Simon and his classmates conducted interviews in their town to find out what TV programme people were watching.

TV programme	Percentage
The news	30%
Sitcom	10%
Game show	5%
Soap opera	55%

Use the data they collected to draw a pie chart.

8 For each of these pie charts, copy and complete the table of data.

a)

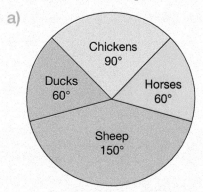

The animals on a farm

Farm animals	Fraction of animals	% animals
Horses		
Sheep		
Ducks		
Chickens		

b)

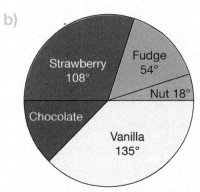

Favourite ice cream flavours

Ice cream flavours	Fraction of flavours	% flavours
Vanilla		
Chocolate		
Strawberry		
Fudge		
Nut		

9 This pie chart shows how much money was spent on health care in a country in 1998.

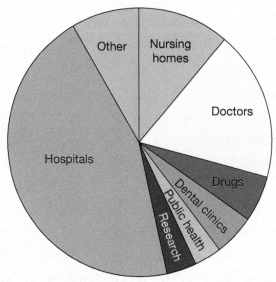

Health care 1998 – how the money is spent
(Total $720 million)

Measure the angle of each sector accurately and then copy and complete the table of data on page 258.

Sector	Angle	Amount of money ($ million)	Percentage
Hospitals			
Doctors			
Nursing homes			
Others			
Drugs			
Dental clinics			
Public health			
Research			

10 This bar chart shows the number of cars of different colours sold in one year.

Copy and complete the table of data values, and then draw an accurate, labelled pie chart for the same data in the bar chart.

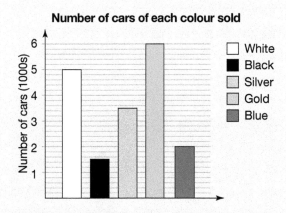

Number of cars of each colour sold

Colour of cars	Number of cars	Fraction	Angle of sector
White			
Black			
Silver			
Gold			
Blue			

11 A group of people were asked which fruit they liked the best – pineapple, mango, melon or peach.
The data from these interviews is shown in the pie chart.

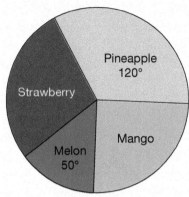

Favourite fruit

a) One-quarter of the people asked said they like mango best.
   Calculate the angle of the 'mango' sector.
b) Calculate the angle of the 'strawberry' sector.
c) If eighteen people said they liked mango best, calculate the total number of people questioned.
d) Calculate the percentage of people who like strawberry the best.

Collecting data through interviews

There are different ways to collect data so that we can analyse it to answer questions.

One way is to ask people. This is called 'conducting an interview'.

Activity

a) Conduct interviews with pupils and teachers in your school. You must choose the question you will ask. Make sure there are at least five different choices/answers.
b) Decide how many pupils and teachers you will interview (must be more that 30).
c) Think about when you will interview them – in the morning before class, or during lunchtime, for example. Remember to thank everyone for helping you to collect the data!
d) Record their answers in the form of a frequency table.
e) Present the data from your interviews as both a histogram and a pie chart.
   Remember to show all the calculations you do to work out the angles of the sectors etc.
f) Think about the data you have collected. What does it tell you? Try to make some conclusions from it.